WITHDRAWN
University of
Illinois Library
at Urbana-Champaign

TAXONOMY OF COMPUTER SCIENCE & ENGINEERING

Compiled by the
AFIPS Taxonomy Committee

Robert L. Ashenhurst
Harold Borko
Sam D. Conte
Patrick Fischer
Bernard A. Galler
George Glaser
Anthony Ralston
(Chairman)
Edwin D. Reilly
Gerard Salton
Jean E. Sammet
Stephen S. Yau

with the help of
over 70 consultants

Printed in the United States of America

Library of Congress Catalog Card No. 79-57474

© 1980 by the American Federation of Information Processing Societies, Inc., Arlington, Va. 22209. All rights reserved. This report or parts thereof may not be reproduced in any form without permission of the publisher.

ACKNOWLEDGEMENTS

This project has been funded by:

Institute for Computer Sciences and Technology of the National Bureau of Standards

Air Force Office of Scientific Research

Office of Naval Research

AFIPS is most appreciative to these agencies for their support.

Copious thanks are also due to Carol Harter and Janet Root who typed the many iterations of this report and maintained the file from which the Taxonomy Tree and Annotated Taxonomy Tree were produced camera-ready for this book.

PREFACE

This document is the result of three years work by a committee of eleven with the aid of over seventy others who either authored part of the *Taxonomy* or reviewed the work of the authors. The result is a study of the structure of a discipline which appears to be unique among the sciences and almost unique among all disciplines.

Why should computer scientists and engineers, whose discipline is the newest among all the scientific and technical disciplines, embark on such a project when their confreres in older, more established disciplines have not done so? The essential reason is lack of understanding of what computer science and engineering is by those outside the discipline. The newness of the field is one reason for this; the other is its very rapid growth, faster, in fact, by any measure than the growth at any time of any other discipline in the sciences or humanities. And, if the pace at the beginning of the fourth decade of the existence of computers is not as frantic as in the first three, it is still so rapid that even those within the discipline are hard pressed to be conversant with anything but a narrow specialty.

But does it matter—except perhaps to our self-esteem—that we are ill-understood or even misunderstood by educated laymen, by government and even by other scientists and engineers? We think it does. Our impact in late twentieth century society is already large and will become immense. If universities still do not always understand the need to provide (informed) instruction in our subject, if governments do not understand our role in their own efficient functioning or in the economics they oversee—to mention only two aspects of the problem—the results can be serious—or worse. This *Taxonomy* will not solve these problems. But as it contributes to increased understanding and heightened perception of what we are and what we do, it will have served a useful role.

For any discipline, but particularly for one changing so rapidly, a taxonomy is only a snapshot, a picture at a particular time. That time for this document is (early) 1979 although, to be fair, almost all of it was produced in 1977 and 1978. If this document proves to be a useful one, then it will need periodic revision. Moreover, the first attempt at a task of this magnitude is surely imperfect in a variety of ways. We would, therefore, like to encourage all readers to communicate to us

—any errors or inaccuracies which you may find (or suspect)
—omissions which you believe are significant
—comments aimed at improving the presentation or annotations.

These should be sent to:

Chairman, Taxonomy Committee
American Federation of Information Processing Societies
1815 North Lynn Street
Arlington, VA 22209

LIST OF CONTRIBUTORS

Taxonomy Committee

Robert L. Ashenhurst
University of Chicago

Harold Borko
UCLA

Sam D. Conte
Purdue University

Patrick Fischer
Vanderbilt University

Bernard A. Galler
University of Michigan

George Glaser
Centigram Corporation

Anthony Ralston
(Chairman)
SUNY at Buffalo

Edwin D. Reilly
SUNY at Albany

Gerard Salton
Cornell University

Jean E. Sammet
IBM Corporation

Stephen S. Yau
Northwestern University

Other Authors

Jon T. Butler
Northwestern University

Billy Claybrook
University of South Carolina

Gerald Engel
Christopher Newport College

Carl Engelman
Mitre Corporation

T. Y. Feng
Wright State University

Aaron Finerman
University of Michigan

Fred H. Harris
University of Chicago

Harold Highland
SUNY at Farmingdale

Laveen Kanal
University of Maryland

Rob Kling
University of California at Irvine

Franklin F. Kuo
University of Hawaii

C. L. Liu
University of Illinois, Champaign

Harold Lorin
IBM Corporation

Joel Moses
M.I.T.

Susan Nycum
Attorney, Chickering & Gregory

Vaughn Pratt
M.I.T.

C. V. Ramamoorthy
University of California at Berkeley

Fred T. Riggs
University of Hawaii

Saul Rosen
Purdue University

Arnold Rosenberg
IBM Corporation

Azriel Rosenfeld
University of Maryland

Paul Roth
U.S. Department of Energy

Stuart Shapiro
SUNY at Buffalo

Ben Shneiderman
University of Maryland

Adrian Stokes
Hatfield Polytechnic, England

Jean-Paul Tremblay
University of Saskatchewan

Andries Van Dam
Brown University

Karl Zinn
University of Michigan

Reviewers and Consultants

John Berg
National Bureau of Standards

Robert Bigelow
Attorney, Bigelow & Saltzberg

Abraham Bookstein
University of Chicago

Frederick M. Brasch
Northwestern University

Fred Brooks
University of North Carolina

Thomas Cain
University of Pittsburgh

Richard Canning
Canning Publications

John Case
SUNY at Buffalo

Steven Coons
University of Colorado, Boulder

David Cooper
University College, London

Fernando Corbató
M.I.T.

Ira Cotton
National Bureau of Standards

Donald Crouch
University of Alabama

Philip Dorn
Dorn Computer Consultants

Marvin Ehlers
Square D Corporation

Donald Frazer
IBM Corporation

John Goodenough
Softech, Inc.

C. C. Gotlieb
University of Toronto

James Griesmer
IBM Corporation

Michael Hammer
M.I.T.

David Hsaio
Ohio State University

Edwin Istvan
National Bureau of Standards

Richard Karhuse
Northwestern University

John Killeen
University of California at Davis

Peter Kirstein
University College, London

Daniel Klassen
Minnesota Educational Computing Consortium

Jon Meads
Intel Corporation

Cleve Moler
University of New Mexico

Abbe Mowshowitz
University of British Columbia

Bary Pollack
University of British Columbia

Udo Pooch
Texas A&M University

Jeffrey Posdamer
SUNY at Buffalo

Keith Price
University of Southern California

Robert Rosin
Bell Laboratories

Sartaj Sahni
University of Minnesota

Robert Sargent
Syracuse University

Alex Shistakov
Lawrence Livermore Laboratory

Susan Solomon
Eastern Washington University

Sargur Srihari
SUNY at Buffalo

Richard Stark
University of New Mexico

Henry Tropp
Humboldt State University, California

Stephen Unger
Columbia University

Terry Winograd
Stanford University

Patrick Winston
M.I.T.

TABLE OF CONTENTS

Acknowledgements	iii
Preface	v
List of Contributors	vii
I. Introduction and Guide to the Taxonomy	1
II. Taxonomy Tree Outline	7
III. Taxonomy Tree	9
IV. Annotated Taxonomy Tree	141
V. Core Terminology	407
VI. Umbrella Terms	411
VII. The Development of the Taxonomy: Philosophy and Technical Issues	415
VIII. Representative Bibliography	423
IX. Abbreviations and Acronyms	427
X. Index	429

I. INTRODUCTION AND GUIDE

If the purpose of producing a taxonomy of computer science and engineering (CS&E) were to be summarized in one word, that word would be *perspective*. The rate of change and progress in CS&E has been so rapid that it has been very difficult to obtain a coherent view of what we are and what we do from outside the discipline and even, sometimes, from within it. If, then, this document, snapshot though it is, can provide a wide-angle view of CS&E, it may help to achieve the perspective of our discipline which is necessary if society is going to use well and wisely the products of CS&E and if the practitioners of the discipline itself are to be useful advisers on applications of and new directions for computer technology.

The starting point in the development of a taxonomy of any subject must be a definition of what that subject encompasses. It seems to us, perhaps parochially, that the boundaries of CS&E are more difficult to ascertain than those of more established disciplines. Partly this results from the rapid growth alluded to above; partly it is because our discipline impinges on areas of more established ones in ways which make it difficult to determine where many topics should be classified. We do not wish to appear to be empire builders gathering into the bosom of CS&E ever larger areas of knowledge and activity. Neither do we wish to appear to be so timid that we do not include areas where, whatever the history, the main contributions are now being made by computer scientists and engineers. The definition which follows is, we believe, a reasonable compromise between these extremes.

Definition of Computer Science and Engineering

Computer science and engineering includes all subject matter normally subsumed under the following rubrics:

    COMPUTER SOFTWARE
    COMPUTER HARDWARE
    COMPUTER SYSTEMS
    THEORY OF COMPUTATION
    THE COMPUTER INDUSTRY
    HISTORY OF COMPUTERS
    EDUCATION IN COMPUTER SCIENCE AND
      ENGINEERING
    LEGAL, MANAGEMENT, PROFESSIONAL
      AND SOCIETAL ASPECTS OF COMPUTING

and, as indicated, portions of the subject matter of

    MATHEMATICS OF COMPUTING
    (namely: much but not all of Numerical Mathematics; some of Automata and Switching Theory; only Random Number Generation and related topics from the general domain of Statistics)

    COMPUTER METHODOLOGIES
    (namely: those methods and techniques with wide applicability, specifically Algebraic Manipulation, Artificial Intelligence, Computer Graphics, Database Management Systems, Image Processing, Information Storage and Retrieval, Pattern Recognition, Simulation, Sorting and Searching)

APPLICATIONS/TECHNIQUES (those traditionally considered part of the discipline which would disappear or be very severely impeded if the computer disappeared; illustrative of these are Business Data Processing, Scientific and Engineering Data Processing, Computer-assisted Instruction, Text (Word) Processing).

Some discussion of this definition may be useful. First, one might not expect to find topics such as the History of Computers or the Legal Aspects of Computing included in a document which attempts to define the structure of a discipline. We took the point of view, however, that an understanding of what CS&E is requires a knowledge of all areas in which computer scientists and engineers work and publish and that, indeed, these topics and other similar ones in the definition above provide a necessary breadth of view about CS&E.

A second decision was to include very little explicitly about the applications of computers to the solution of the myriad problems to which they are now applied. To have been exhaustive here would have meant that the applications part of the *Taxonomy* would have dwarfed the rest of it. Moreover, this portion would have become dated and inaccurate much more rapidly than we hope will be true of the rest of the document. We have, therefore, with some misgivings, because of the inevitably biased point of view which results, restricted our applications section to a few illustrative areas among those where, without computers, the applications would become immensely difficult, if not impossible.

The *Taxonomy*, which is the heart of this document, may be viewed as an elaboration of this definition. It is intended to enable the educated layman to derive some understanding of what we do and, at the same time, to permit those of us in various corners of the discipline to understand better what others in the discipline are concerned with.

Structure of the Report

The remainder of this document consists of the following major sections (the numbers correspond to the Table of Contents in which this section is I.).

II. An outline of the taxonomy tree. This tree contains up to six levels of depth; in this outline we present only the first two levels for the purpose of providing a broad overview.

III. The complete taxonomy tree.

IV. The complete tree again but this time with annotations of terms interleaved with the tree itself. Our aim has been to provide brief definitions or be otherwise descriptive for all terms whose meaning would not be transparent to the educated layman.

V. A list of core terminology in CS&E with definitions. This includes terms in wide use but which, for one reason or another, do not appear explicitly in the tree.

VI. A list of "umbrella" words which, although widely used, have such a broad connotation that they, too, do not appear explicitly in the tree.

VII. An essay which discusses some philosophical and technical issues which we found difficult or contentious.

VIII. A bibliography, keyed to first or second level nodes, containing references—mainly books—to which the user of this report may go for further information on the subject matter of CS&E.

IX. A list of abbreviations and acronyms used in the *Taxonomy*.

X. An alphabetically ordered index of terms which will enable the user of this taxonomy to find the location of terms in the tree or in the lists described in V and VI above.

Structure of the Taxonomy Tree

The *Taxonomy* presented here is in standard indented outline form. In the parlance of CS&E, our structure is a tree with CS&E the label for the root of the tree and Hardware, Computer Systems, Data, Software, Mathematics of Computing, Theory of Computation, Methodologies, Applications/Techniques, and Computing Milieux being the labels for the "nodes" at the ends of the branches from the root. Each level of indentation of the outline corresponds to nodes joined by branches to the next higher level node. The furthest depth of indentation—4 levels in some places, 5 or 6 in others—corresponds to the leaves of the tree. Of course, despite the fact that nominally we have a tree, CS&E can not, any more than any other discipline, be described in such a structurally clean way. In fact, our tree is not really a tree at all but, because of the cross-references (*see* and *see also*), it is really a graph. That is, the structure consists of nodes and edges between pairs of nodes and contains, as a tree may not, cycles (i.e. paths beginning and ending at the same node).

Which is, of course, what would be expected. The interrelationships between Hardware, Software, Data and Computer Systems, between these topics and the Theory of Computation, between Methodologies and Applications etc., assure that only with ample cross-referencing can the structure be wholly perceived.

But we have no doubts that a tree is the appropriate mechanism to use in displaying the structure. The familiarity of the outline form together with the fact that cross-references are the exception and not the rule means that a tree provides more information in a convenient form than any other mode of presentation which we might have chosen.

A few more notes are in order about the way the tree has been constructed:

1. Some of the higher level nodes are not terms in CS&E itself but rather *covering* nodes whose only purpose is to collect lower level concepts under a single heading (e.g., 2.1 Structure-based Systems; 4.2.3 Human involvement).

2. No node has a single offspring (i.e. branch) emanating from it.

3. The order of presentation of the subnodes of a given node generally has some rationale such as:

—from practical to theoretical (or vice versa)
—alphabetical order.

4. A limit of 6 levels of the tree has been rigorously enforced. This was done to keep the total size of the project under control and because no case was presented to the committee of a need for more levels where the result would not have been a *Taxonomy* too detailed to serve our intended purposes.

5. Nodes labeled "Other" have been used rather freely at levels 2, 3, and 4 to indicate that the other subnodes of the parent node are not an exhaustive subdivision of the label of that parent node. However, it is intended that no major concepts be subsumed under "Other."

Guide to the Taxonomy

Users of this *Taxonomy* will come to it with varying orientations and needs. We list here a few of the more obvious ways in which it may be convenient to make use of the *Taxonomy*.

1. Although this document is not, and is not intended to be used as a dictionary or encyclopedia, it will often be useful to look up a term in the Index and then find its place in the annotated tree for a definition of it or information about it and an indication of where it fits into the larger structure.

2. Where the main interest is in obtaining a broad overview of a particular subdisciplinary area, it will often be best to start with the outline of the tree and then proceed to the unannotated tree. (Note that both the unannotated and annotated trees have running heads corresponding to the first or second level nodes in the outline.)

3. The annotated tree will have its major use for those who wish a view in depth of some (relatively narrow) portion of CS&E. As such, it may provide a starting point for a particular study of some area with the bibliography providing pointers to sources for deeper study.

Other Uses of the Taxonomy

In addition to the direct and rather specific uses of this document described in the Guide above, there are a number of more general uses to which the *Taxonomy* may be put. One such might be in the allocation of grant funds to support basic and/or applied research in computer science and engineering. From the structure presented here, it should be possible to understand better the component subareas of various areas of research. The distinctions made, for example, between various kinds of computer architecture (node 1.3) might be helpful in understanding an author's emphasis on research on one particular kind of architecture. In this context, it is worth noting that, if this *Taxonomy* had been generated 10 years ago, the entire area of computer architecture might have been represented in a different form, leading to a different understanding of its role within computer science and engineering. The implication of this is, of course, that a taxonomy like this one must be a living document, changing and adapting as the discipline develops and changes.

Another use which might be made of this *Taxonomy* is in the classification of jobs within a large organization or within governmental civil service. There are many ways to classify positions in industry or government, but usually there is an implicit underlying classification or taxonomy of the field itself. It may well be the case that the difficulty that the federal government and other governmental bodies and large organizations have had in classifying computer science and engineering positions may arise in large measure from a lack of any clear perception of the discipline of CS&E. We hope this *Taxonomy* will be able to provide such perception. Moreover, the structure and relationship between subfields shown here might very well suggest distinctions and/or similarities between otherwise unrelated (or too closely-related) jobs.

Still another application area for this *Taxonomy* may be in the organization of catalogues, review documents, and journals. *Computing Reviews*, for example, is the leading review journal for computer science and engineering publications; it has its own "review categories," developed some years ago. It may be expected that this *Taxonomy* will have some effect on the evolution of the review categories used by *Computing Reviews* and other, related journals.

Finally, one may expect a journalist or other newcomer to the field to find this *Taxonomy* helpful in obtaining an overview and in learning some of the basic terminology. The kinds of distinctions made in creating a taxonomy are precisely those needed to understand arguments and predictions about past history and future growth and emphasis within a discipline.

A document such as this one has, unfortunately, not only its uses but also its potential abuses. The most likely abuse, perhaps, is that, when it conforms to the notions of those with axes to be ground, it will be quoted as revealed truth. It is, of course, nothing of the kind. Even though there has been widespread input from the CS&E community, this *Taxonomy* is, as we indicate elsewhere, inevitably the result of much compromise. Moreover, as the first attempt at the task of classifying the subject matter of CS&E, it must be considered only a first approximation.

It is also possible to abuse our intent by reading more into this *Taxonomy* than was intended. Readers may tend to measure the relative importance of the topics in the *Taxonomy* by such inappropriate means as the space devoted to each first or second level node. Or they may forget that the rapidly changing structure of CS&E requires all use to take into account the creation date of this document. In general, abuse is likely whenever it is forgotten that this *Taxonomy* is and can be nothing more than a snapshot of the discipline taken from one of several possible angles.

Our hope is that we have created a document which will stimulate readers to find unexpected uses but not abuses of this *Taxonomy*. As these new uses are found and become understood, we shall undoubtedly find ways to improve the structure and the presentation of the information herein. We are looking forward to that task.

A final note. Development of this *Taxonomy* has been a humbling task for all of those who have worked on it. There is, inevitably, no single *right way* to classify the subject matter of CS&E. Within the Committee which oversaw this project, between Committee members and authors of portions of the *Taxonomy*, and between authors and reviewers there have been disputes, sometimes heated ones. Not only is it inevitable that our decisions will not satisfy everyone or, perhaps, anyone; it is also inevitable that there will be omissions, possibly some egregious ones, in this document. To a degree, tasks like this one must be undertaken a first time so that they may be got right the next time. We urgently invite constructive criticism from our readers which will inform the next version of this document.

II. TAXONOMY TREE OUTLINE

1. **Hardware**
    1.1 Types of computers
    1.2 Digital computer subsystems
    1.3 Digital computer architecture
    1.4 Input/Output devices
    1.5 Computer circuitry
    1.6 Computer elements
    1.7 Computer hardware reliability

2. **Computer Systems**
    2.1 Structure-based systems
    2.2 Access-based systems
    2.3 Special purpose systems
    2.4 Performance of systems

3. **Data**
    3.1 Data structures
    3.2 Data storage representation
    3.3 Data management
    3.4 Data communications

4. **Software**
    4.1 Tools and techniques
    4.2 Programming systems
    4.3 Data and file organization and management

5. **Mathematics of Computing**
    5.1 Continuous mathematics
    5.2 Discrete mathematics
    5.3 Numerical software and algorithm analysis

6. **Theory of Computation**
    6.1 Switching and automata theory
    6.2 Formal languages
    6.3 Analysis of programs
    6.4 Computer models
    6.5 Complexity of computations
    6.6 Analysis of algorithms

7. **Methodologies**
    7.1 Algebraic manipulation
    7.2 Artificial intelligence
    7.3 Information storage and retrieval
    7.4 Database management systems
    7.5 Image processing
    7.6 Pattern recognition
    7.7 Modeling and simulation
    7.8 Sorting and searching
    7.9 Computer graphics

8. **Applications/Techniques (Illustrative)**
    8.1 Business data processing
    8.2 Scientific and engineering data processing techniques
    8.3 Computer-assisted instruction
    8.4 Text (word) processing

9. **Computing Milieux**
    9.1 The computer industry
    9.2 Education and computing
    9.3 History of computing
    9.4 Legal aspects of computing
    9.5 Management of computing
    9.6 The computing profession
    9.7 Social issues and impacts of computing

III. TAXONOMY TREE

PART III. TAXONOMY TREE
1. HARDWARE

1. <u>Hardware</u>

    1.1   Types of Computers

        1.1.1   Digital computer

            1.1.1.1   Microcomputer
            1.1.1.2   Minicomputer
            1.1.1.3   Medium-scale computer
            1.1.1.4   Large-scale computer
            1.1.1.5   Supercomputer

        1.1.2   Analog computer

        1.1.3   Hybrid computer

            1.1.3.1   Basic patchable hybrid computer
            1.1.3.2   Patchable clocked system
            1.1.3.3   Mini-hybrid computer
            1.1.3.4   Full hybrid computer

    1.2   Digital Computer Subsystems

        1.2.1   Central processing unit (CPU)

            1.2.1.1   Arithmetic-logic unit
            1.2.1.2   Control unit

                1.2.1.2.1   Hardwired
                1.2.1.2.2   Microprogrammed

        1.2.2   Memory

            1.2.2.1   Types

                1.2.2.1.1   Cache memory
                1.2.2.1.2   Main memory
                1.2.2.1.3   Mass memory
                1.2.2.1.4   Buffer memory

            1.2.2.2   Memory access modes

                1.2.2.2.1   Coordinate-addressed access

                    1.2.2.2.1.1   Random access
                    1.2.2.2.1.2   Sequential access
                    1.2.2.2.1.3   Multidimensional access

                1.2.2.2.2   Content-addressed access

        1.2.3   Input/Output (I/O)

            1.2.3.1   Communication path

                1.2.3.1.1   Register-based I/O
                1.2.3.1.2   Memory-based I/O

                    1.2.3.1.2.1   CPU resident controller
                    1.2.3.1.2.2   I/O processor (I/O channel)

            1.2.3.2   Signaling mechanism

                1.2.3.2.1   Programmed I/O

1. HARDWARE

      1.2.3.2.2 Interrupt driven I/O

    1.2.3.3 Input/Output addressing

      1.2.3.3.1 Separate I/O space
      1.2.3.3.2 Memory mapped

1.3 Digital Computer Architecture

  1.3.1 Sequential processor
  1.3.2 Parallel processor

    1.3.2.1 Homogeneous parallel processors

      1.3.2.1.1 Associative processor
      1.3.2.1.2 Ensemble processor
      1.3.2.1.3 Array processor
      1.3.2.1.4 Multiprocessor
      1.3.2.1.5 Vector processor

    1.3.2.2 Heterogeneous parallel processors

      1.3.2.2.1 Pipelined machine
      1.3.2.2.2 Data flow machine
      1.3.2.2.3 Distributed processors (<u>see</u> <u>also</u> 2.1.2.1)
      1.3.2.2.4 Computer network

1.4 Input/Output Devices

  1.4.1 Data entry/retrieval devices

    1.4.1.1 Card reader/punch
    1.4.1.2 Paper tape reader/punch
    1.4.1.3 Printer

      1.4.1.3.1 Line printer
      1.4.1.3.2 Character printer
      1.4.1.3.3 Laser printer

    1.4.1.4 Plotter
    1.4.1.5 Optical scanner
    1.4.1.6 Microfilm recorder
    1.4.1.7 Speech recognizer

  1.4.2 Terminals

    1.4.2.1 Keyboard/printer terminal

      1.4.2.1.1 Teletypes and impact devices
      1.4.2.1.2 Matrix printing devices
      1.4.2.1.3 Non-impact devices

    1.4.2.2 Cathode ray tube (CRT) terminal
    1.4.2.3 Graphic terminal
    1.4.2.4 Intelligent terminal

  1.4.3 Peripheral devices

    1.4.3.1 Magnetic tape device

      1.4.3.1.1 Cassette
      1.4.3.1.2 Cartridge
      1.4.3.1.3 Reel

        1.4.3.2   Magnetic disk and drum devices

                1.4.3.2.1 Floppy disk
                1.4.3.2.2 Cartridge disk
                1.4.3.2.3 Disk pack
                1.4.3.2.4 Drum

        1.4.3.3   Magnetic card devices

        1.4.3.4   Display devices

                1.4.3.4.1 CRT
                1.4.3.4.2 TV
                1.4.3.4.3 Plasma

   1.4.4   I/O processor

        1.4.4.1   Front-end processor
        1.4.4.2   Remote concentrator
        1.4.4.3   Message-switching processor

   1.4.5   Modems

        1.4.5.1   Direct coupler
        1.4.5.2   Acoustic coupler

1.5   Computer Circuitry

   1.5.1   Digital circuitry (logic circuitry)

        1.5.1.1   Combinational circuits

                1.5.1.1.1 Contact circuit
                1.5.1.1.2 Symmetric circuit
                1.5.1.1.3 Threshold circuit

        1.5.1.2   Sequential circuits

                1.5.1.2.1 Mealy model
                1.5.1.2.2 Moore model

        1.5.1.3   Iterative array
        1.5.1.4   Special logic circuits

                1.5.1.4.1 Flip-flop
                1.5.1.4.2 Shift register
                1.5.1.4.3 Counter
                1.5.1.4.4 Comparator
                1.5.1.4.5 Adder
                1.5.1.4.6 Multiplier
                1.5.1.4.7 Universal logic circuit
                1.5.1.4.8 Multiplexor
                1.5.1.4.9 Decoder

   1.5.2   Analog circuitry

        1.5.2.1   Operational amplifier
        1.5.2.2   Integrator
        1.5.2.3   Differentiator

   1.5.3   Hybrid circuitry

        1.5.3.1   Analog-to-digital converter

1. HARDWARE

     1.5.3.2 Digital-to-analog converter
     1.5.3.3 Schmitt trigger

 1.6 Computer Elements

   1.6.1 Semiconductor elements

     1.6.1.1 Bipolar
     1.6.1.2 Field-effect

   1.6.2 Magnetic elements

     1.6.2.1 Magnetic core

       1.6.2.1.1 Single hole
       1.6.2.1.2 Multiaperture

     1.6.2.2 Magnetic thin film

       1.6.2.2.1 Plated wire
       1.6.2.2.2 Flat film

     1.6.2.3 Magnetic bubble device

   1.6.3 Charge-coupled device

   1.6.4 Cryogenic element

   1.6.5 Optical element

   1.6.6 Mechanical element

   1.6.7 Hydraulic element

 1.7 Computer Hardware Reliability

   1.7.1 Measures
   1.7.2 Fault classification

     1.7.2.1 Number of faults

       1.7.2.1.1 Single fault
       1.7.2.1.2 Multiple fault

     1.7.2.2 Types of faults

       1.7.2.2.1 Permanent fault

         1.7.2.2.1.1 Stuck-at fault
         1.7.2.2.1.2 Shorted fault

       1.7.2.2.2 Transient fault

   1.7.3 Methodologies

     1.7.3.1 Element improvement
     1.7.3.2 Component improvement

       1.7.3.2.1 Static redundancies
       1.7.3.2.2 Dynamic redundancies
       1.7.3.2.3 Hybrid redundancies

   1.7.4 Test generation approaches
   1.7.5 Fault simulation

PART III. TAXONOMY TREE
2. COMPUTER SYSTEMS

2. Computer Systems

2.1 Structure-based Systems

2.1.1 Hybrid systems

2.1.2 Multiple processor systems

- 2.1.2.1 Distributed systems (see also 1.3.2.2.3)
- 2.1.2.2 Hierarchical systems
  - 2.1.2.2.1 Tightly-coupled systems
  - 2.1.2.2.2 Loosely-coupled systems
- 2.1.2.3 Shared memory systems
- 2.1.2.4 Master-slave systems

2.1.3 Parallel systems

- 2.1.3.1 Pipeline processors
- 2.1.3.2 Array of processors
- 2.1.3.3 Multiple functional units

2.2 Access-based Systems

- 2.2.1 Batch systems
- 2.2.2 Remote job entry systems
- 2.2.3 Interactive systems
  - 2.2.3.1 Time-sharing systems
  - 2.2.3.2 Dedicated applications systems (see also 2.3.4.6)
- 2.2.4 Communications systems
- 2.2.5 Teleprocessing systems

2.3 Special Purpose Systems

- 2.3.1 Graphics-image processing systems (see also 7.9.2)
- 2.3.2 Adaptive systems
  - 2.3.2.1 Self-organizing systems
  - 2.3.2.2 Fault-tolerant systems
- 2.3.3 I/O and memory systems
  - 2.3.3.1 I/O control systems
    - 2.3.3.1.1 Peripheral subsystems
    - 2.3.3.1.2 Concentrators
  - 2.3.3.2 Hierarchical storage systems
- 2.3.4 Application-based systems
  - 2.3.4.1 Air traffic control
  - 2.3.4.2 Process control
  - 2.3.4.3 Word processing (see also 8.4)
  - 2.3.4.4 Message switching
  - 2.3.4.5 Telephone network control
  - 2.3.4.6 Transaction-based

2. COMPUTER SYSTEMS

                2.3.4.6.1 Reservation
                2.3.4.6.2 Point-of-sale
                2.3.4.6.3 Electronic funds transfer (see also 9.4.8.3)
                2.3.4.6.4 Other

        2.3.4.7 Signal processing
        2.3.4.8 Other

2.4 Performance of Systems

    2.4.1 Attributes

        2.4.1.1 Throughput
        2.4.1.2 Response time

    2.4.2 Assessment of attributes

        2.4.2.1 Selection studies

            2.4.2.1.1 Modeling techniques

                2.4.2.1.1.1 Simulation models (see also 7.7)
                2.4.2.1.1.2 Analytic models

            2.4.2.1.2 Measurement techniques

                2.4.2.1.2.1 Hardware tools
                2.4.2.1.2.2 Software tools
                2.4.2.1.2.3 Firmware tools

        2.4.2.2 Improvement studies
        2.4.2.3 Design studies

PART III. TAXONOMY TREE
3. DATA

3. Data

    3.1 Data Structures

        3.1.1 Primitive

            3.1.1.1 Integer
            3.1.1.2 Real
            3.1.1.3 Logical
            3.1.1.4 Character
            3.1.1.5 String
            3.1.1.6 Pointer
            3.1.1.7 Other

        3.1.2 Composite

            3.1.2.1 Sets
            3.1.2.2 Arrays

                3.1.2.2.1 Indexed access
                3.1.2.2.2 Content access

            3.1.2.3 Linear lists

                3.1.2.3.1 Stack
                3.1.2.3.2 Queue

                    3.1.2.3.2.1 FIFO
                    3.1.2.3.2.2 Deque
                    3.1.2.3.2.3 Priority

                3.1.2.3.3 Other

            3.1.2.4 Nonlinear lists

                3.1.2.4.1 Associative
                3.1.2.4.2 List structures

            3.1.2.5 Trees

                3.1.2.5.1 Binary

                    3.1.2.5.1.1 Heap
                    3.1.2.5.1.2 Height-balanced (AVL)
                    3.1.2.5.1.3 Weight-balanced

                3.1.2.5.2 n-ary

                    3.1.2.5.2.1 Trie
                    3.1.2.5.2.2 Balanced n-ary tree

            3.1.2.6 Graphs (see also 5.2.2)

                3.1.2.6.1 Directed (digraph)

                    3.1.2.6.1.1 Acyclic
                    3.1.2.6.1.2 Other

                3.1.2.6.2 Undirected

            3.1.2.7 Structures

        3.1.3 Logical data base (see also 7.4)

3. DATA

    3.1.3.1 Network
    3.1.3.2 Relational
    3.1.3.3 Hierarchical

 3.2 Data Storage Representation

   3.2.1 Primitive items

    3.2.1.1 Numeric

     3.2.1.1.1 Fixed

      3.2.1.1.1.1 Sign-magnitude
      3.2.1.1.1.2 2's complement
      3.2.1.1.1.3 1's complement

     3.2.1.1.2 Floating

      3.2.1.1.2.1 Excess notation
      3.2.1.1.2.2 Multiple-precision
      3.2.1.1.2.3 Other

    3.2.1.2 Character

     3.2.1.2.1 EBCDIC
     3.2.1.2.2 ASCII

    3.2.1.3 Pointer

     3.2.1.3.1 Absolute
     3.2.1.3.2 Relative

    3.2.1.4 Logical

   3.2.2 Record

    3.2.2.1 Fixed length
    3.2.2.2 Variable length

   3.2.3 File

    3.2.3.1 Internal

     3.2.3.1.1 Sequential
     3.2.3.1.2 Linked

      3.2.3.1.2.1 Singly
      3.2.3.1.2.2 Doubly
      3.2.3.1.2.3 Multilinked

     3.2.3.1.3 Hashed

      3.2.3.1.3.1 Division
      3.2.3.1.3.2 Midsquare
      3.2.3.1.3.3 Folding
      3.2.3.1.3.4 Radix transformation
      3.2.3.1.3.5 Digit analysis
      3.2.3.1.3.6 Piece-wise linear
      3.2.3.1.3.7 Other

     3.2.3.1.4 Other

    3.2.3.2 External

                3.2.3.2.1 Unordered
                3.2.3.2.2 Primary key

                           3.2.3.2.2.1 Sequential
                           3.2.3.2.2.2 Indexed sequential
                           3.2.3.2.2.3 Direct

                3.2.3.2.3 Multi-key

                           3.2.3.2.3.1 Multi-list
                           3.2.3.2.3.2 Inverted list

    3.2.4 Physical data base (see also 7.4)

3.3 Data Management

    3.3.1 Management disciplines

           3.3.1.1 Stack-oriented
           3.3.1.2 First fit
           3.3.1.3 Best fit
           3.3.1.4 Next fit
           3.3.1.5 Buddy system

                    3.3.1.5.1 Binary
                    3.3.1.5.2 Fibonacci

           3.3.1.6 Garbage collection

    3.3.2 Coding

           3.3.2.1 Fixed length

                    3.3.2.1.1 Logical
                    3.3.2.1.2 Binary encoded

           3.3.2.2 Variable length

                    3.3.2.2.1 Huffman (see also 5.2.2.4.5)
                    3.3.2.2.2 Shannon-Fano
                    3.3.2.2.3 Tagged

                               3.3.2.2.3.1 Boundary
                               3.3.2.2.3.2 Description

    3.3.3 Compaction

           3.3.3.1 Run-length coding
           3.3.3.2 Pattern substitution
           3.3.3.3 Differencing

    3.3.4 Encryption

           3.3.4.1 Transposition ciphers
           3.3.4.2 Standard substitution ciphers
           3.3.4.3 Algebraic substitution ciphers
           3.3.4.4 Hybrid systems

3.4 Data Communications

    3.4.1 Data transmission

           3.4.1.1 Transmission media

3. DATA

3.4.1.2 Modulation and demodulation

    3.4.1.2.1 Modulation techniques
    3.4.1.2.2 Modems

        3.4.1.2.2.1 Handshaking procedures
        3.4.1.2.2.2 Modem interface control circuits

3.4.1.3 Multiplexing

    3.4.1.3.1 Space division multiplexing
    3.4.1.3.2 Frequency division multiplexing
    3.4.1.3.3 Time division multiplexing

        3.4.1.3.3.1 Synchronous time division multiplexing
        3.4.1.3.3.2 Asynchronous time division multiplexing
        3.4.1.3.3.3 Contention schemes
        3.4.1.3.3.4 Polling schemes

3.4.1.4 Error control

    3.4.1.4.1 Sources of errors

        3.4.1.4.1.1 Noise
        3.4.1.4.1.2 Distortion, attenuation and phase
        3.4.1.4.1.3 Crosstalk
        3.4.1.4.1.4 Other

    3.4.1.4.2 Equalization
    3.4.1.4.3 Coding for error control

        3.4.1.4.3.1 Error detecting codes
        3.4.1.4.3.2 Error correcting codes

    3.4.1.4.4 Protocols for error control

        3.4.1.4.4.1 Negative acknowledgement
        3.4.1.4.4.2 No error acknowledgement

3.4.2 Data switching

    3.4.2.1 Circuit switching

        3.4.2.1.1 Analog switching systems
        3.4.2.1.2 Time division switching
        3.4.2.1.3 Control signaling

    3.4.2.2 Store and forward switching

        3.4.2.2.1 Message switching
        3.4.2.2.2 Packet switching

3.4.3 Economic, legal and regulatory aspects (see also 9.4, 9.7)

    3.4.3.1 Privacy and security
    3.4.3.2 Computer-communications interface
    3.4.3.3 International standards

PART III. TAXONOMY TREE
4. SOFTWARE

4. Software

4.1 Tools and Techniques

4.1.1 Programming languages

4.1.1.1 Development

- 4.1.1.1.1 Design
- 4.1.1.1.2 Definition techniques
- 4.1.1.1.3 Standardization
- 4.1.1.1.4 Automatic programming

4.1.1.2 Implementation (=Translation)

4.1.1.2.1 Types of translators (=Processors)

- 4.1.1.2.1.1 Compiler
- 4.1.1.2.1.2 Interpreter
- 4.1.1.2.1.3 Mixed
- 4.1.1.2.1.4 Cross-compiler

4.1.1.2.2 Translator writing techniques

- 4.1.1.2.2.1 Syntax-directed
- 4.1.1.2.2.2 Intermediate language
- 4.1.1.2.2.3 Preprocessor
- 4.1.1.2.2.4 Optimization

4.1.1.3 Types (=Classifications)

- 4.1.1.3.1 By application area
- 4.1.1.3.2 Assembly
- 4.1.1.3.3 Macro assembly
- 4.1.1.3.4 High-level
- 4.1.1.3.5 Extensible
- 4.1.1.3.6 Table/questionnaire
- 4.1.1.3.7 Nonprocedural
- 4.1.1.3.8 Data definition
- 4.1.1.3.9 Other

4.1.1.4 Commands (=Statements)

4.1.1.4.1 Control structures (see also 6.3.3)

- 4.1.1.4.1.1 Sequential
- 4.1.1.4.1.2 Loop
- 4.1.1.4.1.3 Conditional
- 4.1.1.4.1.4 Unconditional transfer of control
- 4.1.1.4.1.5 Procedure/subroutine invocation
- 4.1.1.4.1.6 Parallel control structures

- 4.1.1.4.2 Storage management
- 4.1.1.4.3 Input/output
- 4.1.1.4.4 Error detecting/correcting
- 4.1.1.4.5 Non-executable

4.1.1.5 Data types and structures

4.1.1.5.1 Data types (see also 3.1.1)

- 4.1.1.5.1.1 Arithmetic

4. SOFTWARE

                      (see 3.1.1.1 and 3.1.1.2)
             4.1.1.5.1.2 Logical
                      (see 3.1.1.3)
             4.1.1.5.1.3 Character
                      (see 3.1.1.4)
             4.1.1.5.1.4 String
                      (see 3.1.1.5)
             4.1.1.5.1.5 Pointer
                      (see 3.1.1.6)
             4.1.1.5.1.6 Other

          4.1.1.5.2 Data structures (see also 3.1.2)

             4.1.1.5.2.1 Arrays
                      (see 3.1.2.2)
             4.1.1.5.2.2 Lists
                      (see 3.1.2.3 and 3.1.2.4)
             4.1.1.5.2.3 Records
                      (see 3.2.2)
             4.1.1.5.2.4 Files
                      (see 3.2.3)
             4.1.1.5.2.5 Other

          4.1.1.5.3 Abstract data type

  4.1.2 Programming and coding techniques

      4.1.2.1 Subroutine (=Procedure)
      4.1.2.2 Coroutine
      4.1.2.3 Recursion
      4.1.2.4 Reentrant code
      4.1.2.5 Error-related

          4.1.2.5.1 Debugging techniques
          4.1.2.5.2 Restart procedures

      4.1.2.6 Other

  4.1.3 Program design and development

      4.1.3.1 Requirements

          4.1.3.1.1 User
          4.1.3.1.2 Hardware
          4.1.3.1.3 Software

      4.1.3.2 Specification

          4.1.3.2.1 Functional
          4.1.3.2.2 Performance

      4.1.3.3 Design techniques

          4.1.3.3.1 Simulation/modeling (see also 7.7)
          4.1.3.3.2 Modularity
          4.1.3.3.3 Top-down/stepwise refinement
          4.1.3.3.4 Two-dimensional descriptions
          4.1.3.3.5 Other

      4.1.3.4 Documentation (see also 9.5.1.3.4, 9.5.2.2.5 and 9.5.5.4.4)

          4.1.3.4.1 Types

4. SOFTWARE

                4.1.3.4.1.1 Program comments
                4.1.3.4.1.2 Two-dimensional
                4.1.3.4.1.3 Narrative

            4.1.3.4.2 Audiences

                4.1.3.4.2.1 User
                4.1.3.4.2.2 Programmer
                4.1.3.4.2.3 Operator

        4.1.3.5 Effectiveness measurements

            4.1.3.5.1 Testing
            4.1.3.5.2 Software monitors

        4.1.3.6 Maintenance

            4.1.3.6.1 Repairs/corrections
            4.1.3.6.2 Enhancement

        4.1.3.7 Theoretical

            4.1.3.7.1 Program specification
            4.1.3.7.2 Program verification

    4.1.4 Portability

        4.1.4.1 Across machines
        4.1.4.2 Across input/output equipment
        4.1.4.3 Across operating systems
        4.1.4.4 Across translators

4.2 Programming Systems

    4.2.1 Operating systems

        4.2.1.1 Command/job control functions
        4.2.1.2 Time characteristics

            4.2.1.2.1 Real time
            4.2.1.2.2 Interactive (_see also_ 2.2.3)
            4.2.1.2.3 Batch (_see also_ 2.2.2)

        4.2.1.3 Elements to be controlled

            4.2.1.3.1 Input/output
            4.2.1.3.2 Storage allocation

                4.2.1.3.2.1 Physical
                4.2.1.3.2.2 Virtual

            4.2.1.3.3 Parallelism
            4.2.1.3.4 Multiprogramming
            4.2.1.3.5 Multiprocessing (_see also_ 2.1.2)
            4.2.1.3.6 Scheduling of processes
            4.2.1.3.7 Interprocess communication

        4.2.1.4 Protection

            4.2.1.4.1 Privacy
            4.2.1.4.2 Security
            4.2.1.4.3 Error handling

4. SOFTWARE

    4.2.2 Utilities

        4.2.2.1 Linkage editor
        4.2.2.2 Loader
        4.2.2.3 Data transfer
        4.2.2.4 Debugging tools
        4.2.2.5 Diagnostic tools
        4.2.2.6 Other

    4.2.3 Human involvement

        4.2.3.1 Operators (see also 9.5.5.2)

            4.2.3.1.1 Scheduling
            4.2.3.1.2 Error recovery
            4.2.3.1.3 Peripheral manipulation

        4.2.3.2 Data entry (see also 9.5.5.3.1)

            4.2.3.2.1 Verification
            4.2.3.2.2 Correction

4.3 Data and File Organization and Management

    4.3.1 File organization (see also 3.2.3)

        4.3.1.1 Physical

            4.3.1.1.1 Storage media
            4.3.1.1.2 Arrangement on storage media
            4.3.1.1.3 Checking methods (e.g. parity)

        4.3.1.2 Logical

            4.3.1.2.1 File description techniques
            4.3.1.2.2 File identification
            4.3.1.2.3 Record layout
            4.3.1.2.4 Keys of records
            4.3.1.2.5 Indexing and access methods

    4.3.2 Data structures and management

        4.3.2.1 Data description techniques
        4.3.2.2 Types of structures
            (see also 4.1.1.5.2)

        4.3.2.3 Data base management systems (see also 7.4)

            4.3.2.3.1 Data base administrator
            4.3.2.3.2 Data models (e.g. relational)

PART III. TAXONOMY TREE
5. MATHEMATICS OF COMPUTING

5. Mathematics of Computing

5.1 Continuous Mathematics

5.1.1 Zeros of nonlinear equations

- 5.1.1.1 Bisection
- 5.1.1.2 Fixed point iteration
- 5.1.1.3 Secant method
- 5.1.1.4 Newton's method
- 5.1.1.5 Muller's method
- 5.1.1.6 Systems of equations
  - 5.1.1.6.1 Newton-based
  - 5.1.1.6.2 Generalized secant
  - 5.1.1.6.3 Gradient
- 5.1.1.7 Special methods for polynomials
- 5.1.1.8 Global convergence techniques

5.1.2 Numerical integration and differentiation

- 5.1.2.1 Finite difference formulas for differentiation
- 5.1.2.2 Numerical integration-equal intervals
  - 5.1.2.2.1 Trapezoidal rule
  - 5.1.2.2.2 Simpson's rule
  - 5.1.2.2.3 Newton-Cotes formulas
- 5.1.2.3 Euler-Maclaurin formulas
- 5.1.2.4 Romberg integration
- 5.1.2.5 Gaussian quadrature
- 5.1.2.6 Adaptive quadrature

5.1.3 Methods for ordinary differential equations

- 5.1.3.1 One step methods
  - 5.1.3.1.1 Taylor series
  - 5.1.3.1.2 Runge-Kutta
  - 5.1.3.1.3 Runge-Kutta Fehlberg
- 5.1.3.2 Multistep methods
  - 5.1.3.2.1 Predictor-corrector
  - 5.1.3.2.2 Variable-order-variable-step
- 5.1.3.3 Extrapolation methods
- 5.1.3.4 Stability
- 5.1.3.5 Convergence
- 5.1.3.6 Methods for stiff equations
- 5.1.3.7 Boundary value problems
  - 5.1.3.7.1 Shooting methods
  - 5.1.3.7.2 Finite difference methods
  - 5.1.3.7.3 Superposition methods
  - 5.1.3.7.4 Collocation (see also 5.1.5.4)
  - 5.1.3.7.5 Finite element methods (see also 5.1.4.1.2 and 5.1.4.2.5)
  - 5.1.3.7.6 Variational methods
  - 5.1.3.7.7 Invariant imbedding
  - 5.1.3.7.8 Continuation methods

5.1.4 Methods for partial differential equations

    5.1.4.1 Elliptic equations

        5.1.4.1.1 Finite difference methods (see also 5.1.2.1)
        5.1.4.1.2 Finite element methods (see also 5.1.3.7.5 and 5.1.4.2.5)
        5.1.4.1.3 Gauss-Seidel and overrelaxation methods
        5.1.4.1.4 Alternating direction implicit methods
        5.1.4.1.5 Fast Poisson solvers
        5.1.4.1.6 Galerkin's method (see also 5.1.5.5)

    5.1.4.2 Initial value problems (see also 5.1.4.1.1)

        5.1.4.2.1 Finite difference methods
        5.1.4.2.2 Method of lines
        5.1.4.2.3 Method of characteristics
        5.1.4.2.4 Convergence and stability
        5.1.4.2.5 Finite element methods (see also 5.1.3.7.5 and 5.1.4.1.2)

5.1.5 Integral and integro-differential equations

    5.1.5.1 Method of successive approximations
    5.1.5.2 Interpolatory methods
    5.1.5.3 Orthonormal expansions
    5.1.5.4 Collocation method (see also 5.1.3.7.4)
    5.1.5.5 Galerkin's method (see also 5.1.4.1.6)
    5.1.5.6 Methods based on numerical integration

5.1.6 Interpolation and approximation

    5.1.6.1 Interpolation

        5.1.6.1.1 Lagrangian
        5.1.6.1.2 Newton form of the interpolating polynomial
        5.1.6.1.3 Divided differences
        5.1.6.1.4 Hermite
        5.1.6.1.5 Spline (see also 5.1.6.2.6)
        5.1.6.1.6 Trigonometric

    5.1.6.2 Approximation

        5.1.6.2.1 Least squares polynomial
        5.1.6.2.2 Fourier least squares
        5.1.6.2.3 Fast Fourier transform
        5.1.6.2.4 Minimax polynomial
        5.1.6.2.5 Rational
        5.1.6.2.6 Spline (see also 5.1.6.1.5)
        5.1.6.2.7 For elementary functions
        5.1.6.2.8 Data analysis and smoothing

5.1.7 Optimization

    5.1.7.1 Least squares

        5.1.7.1.1 Linear
        5.1.7.1.2 Nonlinear

    5.1.7.2 Minimization of nonlinear functions

        5.1.7.2.1 Newton's method

5. MATHEMATICS OF COMPUTING

           5.1.7.2.2   Quasi-Newton methods
           5.1.7.2.3   Gradient methods

5.2 Discrete Mathematics

    5.2.1   Computer arithmetic

        5.2.1.1   Roundoff error
        5.2.1.2   Floating-point arithmetic
        5.2.1.3   Significant digit arithmetic
        5.2.1.4   Interval arithmetic
        5.2.1.5   Radix number systems

    5.2.2   Graph theory

        5.2.2.1   Basic concepts

           5.2.2.1.1   Directed graphs
           5.2.2.1.2   Undirected graphs
           5.2.2.1.3   Operations on graphs
           5.2.2.1.4   Domination and independence
           5.2.2.1.5   Linear graphs
           5.2.2.1.6   Matching and factors

        5.2.2.2   Paths and circuits

           5.2.2.2.1   Connectivity
           5.2.2.2.2   Eulerian path and circuit
           5.2.2.2.3   Hamiltonian path and circuit
           5.2.2.2.4   The shortest path problem
           5.2.2.2.5   The travelling salesman problem
           5.2.2.2.6   Transitive closure

        5.2.2.3   Planar graphs

           5.2.2.3.1   Euler's formula
           5.2.2.3.2   Kuratowski's theorem
           5.2.2.3.3   Dual graphs
           5.2.2.3.4   Genus, thickness, coarseness, crossing number
           5.2.2.3.5   Colorability and chromatic number
           5.2.2.3.6   The four-color theorem

        5.2.2.4   Trees

           5.2.2.4.1   Characterization of trees
           5.2.2.4.2   Free trees and centroids
           5.2.2.4.3   Enumeration of trees
           5.2.2.4.4   Minimum spanning trees
           5.2.2.4.5   Huffman trees (<u>see</u> <u>also</u> 3.3.2.2.1)
           5.2.2.4.6   Search trees
           5.2.2.4.7   Cut-sets
           5.2.2.4.8   Network flow problems

    5.2.3   Combinatorics

        5.2.3.1   Permutations and combinations

           5.2.3.1.1   Permutations
           5.2.3.1.2   Combinations
           5.2.3.1.3   Binomial coefficients and Stirling numbers
           5.2.3.1.4   Generation of permutations and combinations

5. MATHEMATICS OF COMPUTING

- 5.2.3.2 Recurrences
  - 5.2.3.2.1 Linear recurrences
  - 5.2.3.2.2 Nonlinear recurrences

- 5.2.3.3 Generating functions
  - 5.2.3.3.1 Ordinary generating functions
  - 5.2.3.3.2 Exponential generating functions

- 5.2.3.4 Principle of inclusion and exclusion
  - 5.2.3.4.1 General formula
  - 5.2.3.4.2 Mobius inversion

- 5.2.3.5 Polya's theory of counting
- 5.2.3.6 Block designs
- 5.2.3.7 Ramsey theory

5.2.4 Linear equations and linear algebra

- 5.2.4.1 Direct methods for linear systems
  - 5.2.4.1.1 Gaussian elimination (see also 5.2.4.6.2)
  - 5.2.4.1.2 Gauss-Jordan reduction
  - 5.2.4.1.3 Factorization methods
  - 5.2.4.1.4 Sparse matrix methods
  - 5.2.4.1.5 Methods for inverses and pseudoinverses

- 5.2.4.2 Conditioning and error analysis
- 5.2.4.3 Iterative refinement
- 5.2.4.4 Iterative methods
  - 5.2.4.4.1 Jacobi
  - 5.2.4.4.2 Gauss-Seidel
  - 5.2.4.4.3 Successive overrelaxation
  - 5.2.4.4.4 Conjugate gradient
  - 5.2.4.4.5 Other

- 5.2.4.5 Eigenvalues
  - 5.2.4.5.1 Power iteration
  - 5.2.4.5.2 Jacobi method
  - 5.2.4.5.3 Householder's and Givens' method
  - 5.2.4.5.4 Q-R method
  - 5.2.4.5.5 Lanczos' algorithm
  - 5.2.4.5.6 Singular value decomposition

- 5.2.4.6 Determinants
  - 5.2.4.6.1 Expansion by cofactors
  - 5.2.4.6.2 By Gaussian elimination (see also 5.2.4.1.1)

5.2.5 Mathematical programming

- 5.2.5.1 Linear programming
  - 5.2.5.1.1 Simplex method
  - 5.2.5.1.2 Other

- 5.2.5.2 Nonlinear programming
- 5.2.5.3 Integer programming

5. MATHEMATICS OF COMPUTING

5.2.6 Mathematical statistics and probability

    5.2.6.1 Random number generators

        5.2.6.1.1 Linear congruential generators
        5.2.6.1.2 Normal random numbers
        5.2.6.1.3 Other frequency distributions

    5.2.6.2 Monte Carlo methods

        5.2.6.2.1 For integration in multidimensional space
        5.2.6.2.2 For integral equations

    5.2.6.3 Probability distributions

5.3 Numerical Software and Algorithm Analysis

    5.3.1 Algorithm selection factors

        5.3.1.1 Rates of convergence
        5.3.1.2 Efficiency
        5.3.1.3 Error analysis and stability
        5.3.1.4 Memory constraints
        5.3.1.5 Computer architecture constraints

    5.3.2 Numerical software

        5.3.2.1 Portability constraints (see also 4.1.4)
        5.3.2.2 Effect of languages and compilers
        5.3.2.3 Testing and certification
        5.3.2.4 Program verification
        5.3.2.5 Reliability and robustness
        5.3.2.6 Problem-oriented languages and systems

PART III. TAXONOMY TREE
6. THEORY OF COMPUTATION

6. Theory of Computation

6.1 Switching and Automata Theory

6.1.1 Switching theory

- 6.1.1.1 Boolean algebras
- 6.1.1.2 Synthesis of combinational circuits

6.1.2 Sequential machines

- 6.1.2.1 Input/output conventions
- 6.1.2.2 State reduction
- 6.1.2.3 Identification experiments
- 6.1.2.4 Information lossless machines
- 6.1.2.5 Neural nets

6.1.3 Finite-state acceptors

- 6.1.3.1 Closure properties
- 6.1.3.2 Nondeterministic machines
- 6.1.3.3 Regular expressions
- 6.1.3.4 Decision properties

6.1.4 Generalized finite-state machines

- 6.1.4.1 Two-way automata
- 6.1.4.2 Multitape automata
- 6.1.4.3 Probabilistic automata
- 6.1.4.4 Tree automata

6.1.5 Infinite-state machines

- 6.1.5.1 Counter machines
- 6.1.5.2 Pushdown automata
- 6.1.5.3 Stack automata
- 6.1.5.4 Turing machines (see also 6.4.1)
- 6.1.5.5 Iterative arrays of finite automata (see also 6.4.3)

6.2 Formal Languages

6.2.1 The Chomsky hierarchy

- 6.2.1.1 Unrestricted (Type 0) grammars
- 6.2.1.2 Context-sensitive (Type 1) grammars
- 6.2.1.3 Context-free (Type 2) grammars
- 6.2.1.4 One-sided linear (Type 3) grammars

6.2.2 One-sided linear languages

- 6.2.2.1 Characterization by finite automata
- 6.2.2.2 Characterization by regular expressions
- 6.2.2.3 Subclasses of the finite-state languages

6.2.3 Context-free languages and grammars

- 6.2.3.1 Characterization by nondeterministic pushdown automata
- 6.2.3.2 Ambiguous grammars and inherently ambiguous languages
- 6.2.3.3 Canonical forms
- 6.2.3.4 Closure properties
- 6.2.3.5 Decision properties
- 6.2.3.6 Algebraic characterizations

6. THEORY OF COMPUTATION

- 6.2.4 Parsing and recognition of context-free languages
    - 6.2.4.1 Special forms of grammars
    - 6.2.4.2 General purpose parsing algorithms
    - 6.2.4.3 Complexity of parsing and recognition
- 6.2.5 Restricted context-free languages
    - 6.2.5.1 Deterministic context-free languages
    - 6.2.5.2 Linear and metalinear languages
- 6.2.6 Context-sensitive languages
    - 6.2.6.1 Characterization by linear bounded automata
    - 6.2.6.2 Restricted context-sensitive languages
- 6.2.7 Departures from the Chomsky paradigm
    - 6.2.7.1 Generalized rewriting systems
    - 6.2.7.2 Abstract families of languages

6.3 Analysis of Programs

- 6.3.1 Program syntax
    - 6.3.1.1 Concrete syntax
    - 6.3.1.2 Abstract syntax
- 6.3.2 Program semantics
    - 6.3.2.1 Operational semantics
        - 6.3.2.1.1 SECD machine
        - 6.3.2.1.2 Vienna Definition Language
        - 6.3.2.1.3 Lambda-calculus models
        - 6.3.2.1.4 Combinatory logic models
        - 6.3.2.1.5 Petri nets
    - 6.3.2.2 Denotational semantics
        - 6.3.2.2.1 Command sequences
        - 6.3.2.2.2 State trajectories
        - 6.3.2.2.3 Binary relations
        - 6.3.2.2.4 Lattices
        - 6.3.2.2.5 Complete partial orders
        - 6.3.2.2.6 Varieties, or equational classes of algebras
    - 6.3.2.3 Axiomatic semantics
- 6.3.3 Program constructs
    - 6.3.3.1 Assignment
        - 6.3.3.1.1 Unscoped assignment
            - 6.3.3.1.1.1 Simple assignment
            - 6.3.3.1.1.2 Subscripted assignment
            - 6.3.3.1.1.3 Record assignment
        - 6.3.3.1.2 Scoped assignment
            - 6.3.3.1.2.1 Parameter mechanisms
            - 6.3.3.1.2.2 Lexical scoping

6.3.3.1.2.3 Fluid (dynamic) scoping

6.3.3.2 Control (see also 4.1.1.4)

6.3.3.2.1 Sequential control

6.3.3.2.1.1 Conditional statements
6.3.3.2.1.2 Iteration
6.3.3.2.1.3 Procedure calls
6.3.3.2.1.4 Recursion

6.3.3.2.2 Parallel control

6.3.3.2.2.1 Serializable concurrency
6.3.3.2.2.2 Asynchronous concurrency

6.3.4 Program logic

6.3.4.1 Flowchart logics
6.3.4.2 Partial correctness logics
6.3.4.3 Termination logics
6.3.4.4 Algorithmic logics
6.3.4.5 Modal logics

6.3.5 Program optimization

6.3.5.1 Flow analysis
6.3.5.2 Recursion removal

6.3.6 Program schematology

6.3.6.1 Comparative expressiveness
6.3.6.2 Decision problems

6.4 Computer Models

6.4.1 Turing machines (see also 6.1.5.4)

6.4.1.1 One-tape machines

6.4.1.1.1 Turing's original model
6.4.1.1.2 Post's variant
6.4.1.1.3 Wang's variant

6.4.1.2 Multitape machines

6.4.1.2.1 Ordinary tapes
6.4.1.2.2 Multihead tapes
6.4.1.2.3 Multidimensional tapes

6.4.2 Random-access machines

6.4.2.1 Shepherdson-Sturgis machines
6.4.2.2 Loop machines (see also 6.3.3.2.1)
6.4.2.3 Variants

6.4.2.3.1 Indirect addressing
6.4.2.3.2 Bound on register contents
6.4.2.3.3 Larger instruction sets

6.4.3 Iterative arrays (see also 6.1.5.5)

6.4.3.1 One-dimensional arrays

6. THEORY OF COMPUTATION

                6.4.3.1.1 Serial I/O at end of array
                6.4.3.1.2 Parallel I/O to all elements
                6.4.3.1.3 Common bus

        6.4.3.2 Higher dimensional arrays
        6.4.3.3 Pattern reproduction in two-dimensional arrays

6.5 Complexity of Computations

    6.5.1 Time complexity

        6.5.1.1 Real time
        6.5.1.2 Linear time
        6.5.1.3 Polynomial time (see also 6.6.3.1)
        6.5.1.4 Exponential time

    6.5.2 Space complexity

        6.5.2.1 Linear space
        6.5.2.2 Polynomial space
        6.5.2.3 Exponential space
        6.5.2.4 Time vs. space

    6.5.3 Other measures of complexity

        6.5.3.1 Tape reversals
        6.5.3.2 Conditional transfers
        6.5.3.3 Nondeterministic steps
        6.5.3.4 Circuit complexity

    6.5.4 Machine independent complexity theory

        6.5.4.1 Basic axioms
        6.5.4.2 Speed-up properties
        6.5.4.3 Honest complexity classes

6.6 Analysis of Algorithms

    6.6.1 Complexity of numerical problems

        6.6.1.1 Fast Fourier transforms
        6.6.1.2 Evaluation of polynomials

            6.6.1.2.1 Horner's rule
            6.6.1.2.2 Preconditioning

        6.6.1.3 Bilinear forms
        6.6.1.4 Matrix multiplication

            6.6.1.4.1 Reducing arithmetic operations
            6.6.1.4.2 Reducing page faults

        6.6.1.5 Analytic complexity
        6.6.1.6 Parallel algorithms

    6.6.2 Complexity of nonnumeric and combinatorial problems

        6.6.2.1 Searching (see also 7.8.2)
        6.6.2.2 Insertion/deletion
        6.6.2.3 Sorting (see also 7.8.1)
        6.6.2.4 Graph algorithms (see also 5.2.2)

    6.6.3 Apparently hard problems

6.6.3.1 Polynomial time reducibility

    6.6.3.1.1 Cook reducibility
    6.6.3.1.2 Karp reducibility

6.6.3.2 NP-complete problems

    6.6.3.2.1 Satisfiability of a Boolean expression
    6.6.3.2.2 Existence of a k-clique
    6.6.3.2.3 Existence of a Hamiltonian circuit
    6.6.3.2.4 Other equivalent problems

6.6.4 Inherently hard problems

    6.6.4.1 Equivalence of regular expressions
    6.6.4.2 Presburger arithmetic

6.6.5 General unsolvability results

    6.6.5.1 Halting problem
    6.6.5.2 Post correspondence problem
    6.6.5.3 Other equivalent unsolvable problems
    6.6.5.4 Higher degrees of unsolvability

PART III. TAXONOMY TREE
7. METHODOLOGIES

PART III. TAXONOMY TREE
7. METHODOLOGIES
7.1 ALGEBRAIC MANIPULATION

7.1 Algebraic Manipulation

7.1.1 Expressions and their representation

7.1.1.1 Data structures
7.1.1.2 Types of expressions

- 7.1.1.2.1 Polynomials
- 7.1.1.2.2 Rational functions
- 7.1.1.2.3 Algebraic functions
- 7.1.1.2.4 Transcendental functions
- 7.1.1.2.5 Power series
- 7.1.1.2.6 Composite objects
- 7.1.1.2.7 Other

7.1.1.3 Representations of expressions

- 7.1.1.3.1 Polynomial (<u>see</u> <u>also</u> 7.1.1.2.1)
    - 7.1.1.3.1.1 Distributed
    - 7.1.1.3.1.2 Recursive
    - 7.1.1.3.1.3 Factored
- 7.1.1.3.2 General representations

7.1.1.4 Numerical coefficient domains

- 7.1.1.4.1 Infinite precision integers
- 7.1.1.4.2 Rational numbers
- 7.1.1.4.3 Algebraic numbers
- 7.1.1.4.4 Floating-point numbers
- 7.1.1.4.5 Variable precision floating-point numbers

7.1.2 Algorithms

7.1.2.1 Operations on polynomials, rational functions and coefficient domains

- 7.1.2.1.1 Basic arithmetic operations
- 7.1.2.1.2 Greatest common divisor algorithm
- 7.1.2.1.3 Factorization
- 7.1.2.1.4 Determinants and solutions of linear equations
- 7.1.2.1.5 Resultants
- 7.1.2.1.6 Other

7.1.2.2 Simplification

- 7.1.2.2.1 Canonical forms
- 7.1.2.2.2 Algebraic independence
- 7.1.2.2.3 Pattern matching and rule systems
- 7.1.2.2.4 Decidability results
- 7.1.2.2.5 Other

7.1.2.3 Differentiation
7.1.2.4 Limits
7.1.2.5 Integration

- 7.1.2.5.1 Indefinite integration
    - 7.1.2.5.1.1 Rational functions
    - 7.1.2.5.1.2 Exponential and logarithmic functions

7. METHODOLOGIES

                      7.1.2.5.1.3 Algebraic function extensions (<u>see also</u> 7.1.1.2.3)
                      7.1.2.5.1.4 Special functions

            7.1.2.5.2 Definite integration
            7.1.2.5.3 Ordinary differential equations

    7.1.2.6 Summation
    7.1.2.7 Analysis of algorithms

        7.1.2.7.1 Worst-case analysis
        7.1.2.7.2 "Fast" algorithms
        7.1.2.7.3 Exact analysis
        7.1.2.7.4 Probabilistic analysis
        7.1.2.7.5 Average computing time analysis
        7.1.2.7.6 Other

    7.1.2.8 Other

7.1.3 Languages and systems

    7.1.3.1 Type of system
    7.1.3.2 Mode of operation

        7.1.3.2.1 Interactive
        7.1.3.2.2 Batch

    7.1.3.3 Languages

        7.1.3.3.1 Extension of existing languages
        7.1.3.3.2 Procedural languages
        7.1.3.3.3 Nonprocedural languages

    7.1.3.4 Evaluation of expressions

        7.1.3.4.1 Evaluation by substitution
        7.1.3.4.2 Markov algorithms
        7.1.3.4.3 Lambda-calculus evaluation

    7.1.3.5 Input/output facilities

        7.1.3.5.1 Automatic two-dimensional display of expressions
        7.1.3.5.2 Output formatting
        7.1.3.5.3 Handwritten input of expressions

    7.1.3.6 Interface to numerical routines
    7.1.3.7 Specialized processors
    7.1.3.8 Specialized consoles

7.1.4 Applications

    7.1.4.1 Physics

        7.1.4.1.1 General relativity
        7.1.4.1.2 High energy
        7.1.4.1.3 Plasma physics
        7.1.4.1.4 Other

    7.1.4.2 Mathematics

        7.1.4.2.1 Group theory
        7.1.4.2.2 Hydrodynamics

                7.1.4.2.3 Other

    7.1.4.3 Computer science

                7.1.4.3.1 Analysis of algorithms
                7.1.4.3.2 Program verification
                7.1.4.3.3 Numerical analysis

                            7.1.4.3.3.1 Error analysis
                            7.1.4.3.3.2 Partial differential equations
                            7.1.4.3.3.3 Other

7.1.4.4 Astronomy
7.1.4.5 Engineering
7.1.4.6 Education
7.1.4.7 Other

PART III. TAXONOMY TREE
7. METHODOLOGIES
7.2 ARTIFICIAL INTELLIGENCE

7.2 Artificial Intelligence

7.2.1 Automatic programming (see also 4.1.1.1.4)

- 7.2.1.1 Program construction
- 7.2.1.2 Program verification
- 7.2.1.3 Programming assistants

7.2.2 Deduction

- 7.2.2.1 Direct methods
- 7.2.2.2 Refutation methods
- 7.2.2.3 Informal methods
- 7.2.2.4 Non-standard logics

7.2.3 Learning

- 7.2.3.1 Concept formation
- 7.2.3.2 Grammatical inference
- 7.2.3.3 Inductive inference
- 7.2.3.4 Parameter weighting

7.2.4 Natural language processing

- 7.2.4.1 Speech understanding
- 7.2.4.2 Parsing (see also 6.2.4 and 7.3.1.4)
- 7.2.4.3 Discourse analysis
- 7.2.4.4 Sentence generation
- 7.2.4.5 Discourse generation
- 7.2.4.6 Speech production

7.2.5 Problem solving

- 7.2.5.1 Game playing
- 7.2.5.2 Heuristic search
- 7.2.5.3 Planning

7.2.6 Representation of knowledge (see also 3.1.2)

- 7.2.6.1 Network representations (see also 7.3.1.4.6)
- 7.2.6.2 Predicate calculus representations
- 7.2.6.3 Procedural representations

7.2.7 Robotics

- 7.2.7.1 Exploration robots
- 7.2.7.2 Manipulation robots

7.2.8 Vision

- 7.2.8.1 Two-dimensional objects and line drawings (see also 7.6)
- 7.2.8.2 Real-world objects and scenes (see also 7.5.8)
- 7.2.8.3 Motion

7.2.9 Specialized systems

- 7.2.9.1 Chemistry systems
- 7.2.9.2 Intelligent computer-assisted instruction (see also 8.3)
- 7.2.9.3 Medical systems
- 7.2.9.4 Other

7. METHODOLOGIES

7.2.10 Software

    7.2.10.1 Control structures (<u>see</u> <u>also</u> 4.1.1.4.1)

        7.2.10.1.1 Hierarchical

            7.2.10.1.1.1 Recursion
            7.2.10.1.1.2 Backtracking

        7.2.10.1.2 Heterarchical

            7.2.10.1.2.1 Data-directed
            7.2.10.1.2.2 Communicating coroutines

    7.2.10.2 Languages

PART III. TAXONOMY TREE
7. METHODOLOGIES
7.3 INFORMATION STORAGE AND RETRIEVAL

7.3 Information Storage and Retrieval

7.3.1 Information analysis

7.3.1.1 Indexing methods

- 7.3.1.1.1 Term extraction
- 7.3.1.1.2 Term weighting
- 7.3.1.1.3 Word truncation
- 7.3.1.1.4 Thesaurus methods (see 7.3.1.3.2)
- 7.3.1.1.5 Phrase formation
- 7.3.1.1.6 Term associations
- 7.3.1.1.7 Citation indexing
- 7.3.1.1.8 KWIC indexing (see 7.3.1.3.4)
- 7.3.1.1.9 Indexing language construction

7.3.1.2 Abstracting procedures

- 7.3.1.2.1 Sentence extraction
- 7.3.1.2.2 Sentence scoring
- 7.3.1.2.3 Coherence criteria

7.3.1.3 Dictionaries and thesauruses

- 7.3.1.3.1 Word stem dictionary
- 7.3.1.3.2 Synonym dictionary (thesaurus)
- 7.3.1.3.3 Phrase dictionary
- 7.3.1.3.4 Keyword-in-context (KWIC) list
- 7.3.1.3.5 Concept hierarchy
- 7.3.1.3.6 Concordance utilization

7.3.1.4 Linguistic processing

- 7.3.1.4.1 Phrase structure grammar
- 7.3.1.4.2 String grammar
- 7.3.1.4.3 Augmented transition network
- 7.3.1.4.4 Transformational grammar (see also 6.2.1)
- 7.3.1.4.5 Case grammar
- 7.3.1.4.6 Semantic network (see also 7.2.6.1)
- 7.3.1.4.7 Combined syntactic-semantic systems

7.3.2 Information storage

7.3.2.1 Storage organization

- 7.3.2.1.1 Sequential file organization
- 7.3.2.1.2 Chained (multilist) files
- 7.3.2.1.3 Inverted files
- 7.3.2.1.4 Scatter storage files
- 7.3.2.1.5 Clustered files

7.3.2.2 Record classification

- 7.3.2.2.1 Hierarchical grouping methods
- 7.3.2.2.2 Iterative partitioning
- 7.3.2.2.3 Fast single-pass clustering

7.3.2.3 Collection control

- 7.3.2.3.1 Collection growth
- 7.3.2.3.2 Collection retirement

7. METHODOLOGIES

                7.3.2.3.3 Cooperative collection development
                7.3.2.3.4 Bibliometrics

  7.3.3 Information search and retrieval

        7.3.3.1 Query formulation

                7.3.3.1.1 Controlled vocabulary
                7.3.3.1.2 Natural language vocabulary
                7.3.3.1.3 Boolean query formulation
                7.3.3.1.4 Term vector formulation

        7.3.3.2 Search process (see 7.8.2)
        7.3.3.3 Retrieval process

                7.3.3.3.1 Association and correlation coefficients
                7.3.3.3.2 Query-record similarity
                7.3.3.3.3 Record-record similarity

        7.3.3.4 Interactive query processing

                7.3.3.4.1 Vocabulary display
                7.3.3.4.2 Document display
                7.3.3.4.3 Multistep query refinement
                7.3.3.4.4 Automatic feedback process

  7.3.4 Retrieval evaluation

        7.3.4.1 Efficiency

                7.3.4.1.1 Cost-based measures
                7.3.4.1.2 Time-based measures

        7.3.4.2 Effectiveness

                7.3.4.2.1 Recall
                7.3.4.2.2 Precision
                7.3.4.2.3 Single-valued measures

  7.3.5 Retrieval models

        7.3.5.1 Set-theoretic models
        7.3.5.2 Probabilistic models
        7.3.5.3 Decision theoretic models
        7.3.5.4 Information theoretic models

  7.3.6 Information systems

        7.3.6.1 Storage and retrieval systems

                7.3.6.1.1 Reference retrieval systems
                7.3.6.1.2 Question-answering (fact retrieval) systems
                7.3.6.1.3 Current awareness (SDI) systems
                7.3.6.1.4 Structured data base systems (see also 7.4)
                7.3.6.1.5 Integrated information systems
                7.3.6.1.6 Information network

        7.3.6.2 Auxiliary information systems

                7.3.6.2.1 Source data automation

7.3 INFORMATION STORAGE AND RETRIEVAL

```
7.3.6.2.2   Text editing (see 8.4.1)
7.3.6.2.3   Automated publication process
7.3.6.2.4   Automated office systems (see also 8.4.3)
```

PART III. TAXONOMY TREE
7. METHODOLOGIES
7.4 DATABASE MANAGEMENT SYSTEMS

7.4 Database Management Systems

7.4.1 Theory

- 7.4.1.1 Data semantics
- 7.4.1.2 Normalization
- 7.4.1.3 Conceptual schema
- 7.4.1.4 Data model theory
- 7.4.1.5 Formal mathematical descriptions

7.4.2 Logical design

- 7.4.2.1 Data models
  - 7.4.2.1.1 Hierarchical
  - 7.4.2.1.2 Network
  - 7.4.2.1.3 Relational
  - 7.4.2.1.4 Set-theoretic
- 7.4.2.2 Syntactic representation
- 7.4.2.3 Semantic representation
- 7.4.2.4 Data dictionary/directory
- 7.4.2.5 Data quality and integrity
- 7.4.2.6 Data independence

7.4.3 Accessing methods

- 7.4.3.1 Index organization and searching techniques
- 7.4.3.2 Indexed access
- 7.4.3.3 Direct access
- 7.4.3.4 Link and selector access

7.4.4 Language interfaces

- 7.4.4.1 Data description languages
  - 7.4.4.1.1 Schema
  - 7.4.4.1.2 Subschema
  - 7.4.4.1.3 Stored data
  - 7.4.4.1.4 Mapping between levels
- 7.4.4.2 Data manipulation languages
  - 7.4.4.2.1 Self-contained
  - 7.4.4.2.2 Host-embedded
- 7.4.4.3 Query languages
  - 7.4.4.3.1 Keyword
  - 7.4.4.3.2 Two-dimensional
  - 7.4.4.3.3 Specification
  - 7.4.4.3.4 Procedural
  - 7.4.4.3.5 Menu
  - 7.4.4.3.6 Natural language (see also 7.2.4)
- 7.4.4.4 Protection specification languages
- 7.4.4.5 Implementation of language interfaces
  - 7.4.4.5.1 Integrating data base constructs in high level languages
  - 7.4.4.5.2 Embedding data manipulation languages in high level languages

7. METHODOLOGIES

                7.4.4.5.3 Linking programming languages to database management systems

    7.4.5 Physical design

        7.4.5.1 Software architecture
        7.4.5.2 File management

            7.4.5.2.1 Access strategies
            7.4.5.2.2 Storage strategies
            7.4.5.2.3 Buffer management

        7.4.5.3 Concurrency
        7.4.5.4 Dynamic restructuring
        7.4.5.5 Data translation
        7.4.5.6 Modeling and performance evaluation
        7.4.5.7 Data integrity
        7.4.5.8 Database recovery
        7.4.5.9 Security protection (see also 9.5.5.1)
        7.4.5.10 Hardware and machine architecture

            7.4.5.10.1 Backend computer
            7.4.5.10.2 Database machine
            7.4.5.10.3 Associative machine

    7.4.6 Management of database systems

        7.4.6.1 Policy and legal aspects of security (see also 9.4.3 and 9.5.5.1)
        7.4.6.2 Database administrator (see also 9.6.1.6)
        7.4.6.3 Centralization/decentralization (see also 9.5.6.1.1)
        7.4.6.4 Data auditing (see also 9.5.1.4)
        7.4.6.5 Human factors

PART III. TAXONOMY TREE
7. METHODOLOGIES
7.5 IMAGE PROCESSING

7.5 Image Processing (see also 7.9.3)

7.5.1 Digitization

7.5.1.1 Sampling
7.5.1.2 Quantization

7.5.2 Compression (coding) (see also 3.3.2 and 3.3.3)

7.5.2.1 Efficient encoding

  7.5.2.1.1 Shannon-Fano-Huffman
  7.5.2.1.2 Run length
  7.5.2.1.3 Contour

7.5.2.2 Predictive coding
7.5.2.3 Transform coding
7.5.2.4 Adaptive coding
7.5.2.5 Interframe coding

7.5.3 Enhancement

7.5.3.1 Grayscale manipulation
7.5.3.2 Geometric correction
7.5.3.3 Smoothing

  7.5.3.3.1 Noise cleaning
  7.5.3.3.2 Averaging

7.5.3.4 Sharpening, deblurring

  7.5.3.4.1 High-emphasis filtering
  7.5.3.4.2 Unsharp masking, Laplacian processing

7.5.4 Restoration

7.5.4.1 Inverse filtering
7.5.4.2 Wiener filtering
7.5.4.3 Pseudoinverse restoration (see also 5.2.4.1.5)
7.5.4.4 Recursive (Kalman) filtering

7.5.5 Reconstruction (from projections)

7.5.5.1 Summation methods
7.5.5.2 Transform methods
7.5.5.3 Series expansion methods

7.5.6 Segmentation

7.5.6.1 Thresholding
7.5.6.2 Edge detection
7.5.6.3 Matching
7.5.6.4 Line detection
7.5.6.5 Tracking
7.5.6.6 Region growing, partitioning
7.5.6.7 Shrinking, thinning

7.5.7 Feature measurement

7.5.7.1 Geometrical properties, shape
7.5.7.2 Textural properties
7.5.7.3 Moments
7.5.7.4 Projections, cross-sections

7. METHODOLOGIES

                7.5.7.5 Invariants

- 7.5.8 Scene analysis (see also 7.2.8.2)

    - 7.5.8.1 From single images
    - 7.5.8.2 Use of auxiliary information

- 7.5.9 Applications (see also 7.6.3)

    - 7.5.9.1 Document processing
    - 7.5.9.2 Medicine and biology
    - 7.5.9.3 High-energy physics
    - 7.5.9.4 Industrial automation
    - 7.5.9.5 Remote sensing
    - 7.5.9.6 Reconnaissance
    - 7.5.9.7 Forensic sciences
    - 7.5.9.8 Line drawing processing (see also 7.6.3.2)

PART III. TAXONOMY TREE
7. METHODOLOGIES
7.6 PATTERN RECOGNITION

7.6 Pattern Recognition

7.6.1 Models

7.6.1.1 Classificatory models

7.6.1.1.1 Deterministic models
7.6.1.1.2 Stochastic models

    7.6.1.1.2.1 Statistical feature extraction
    7.6.1.1.2.2 Statistical classification and discrimination
    7.6.1.1.2.3 State-space models

7.6.1.1.3 Fuzzy set model

7.6.1.2 Structural models

7.6.1.2.1 Segmentation and primitive identification
7.6.1.2.2 Formal models for structural description and generation
7.6.1.2.3 Integrated segmentation and structural description

    7.6.1.2.3.1 A priori knowledge representations
    7.6.1.2.3.2 Data driven, non-canonical parsers

7.6.2 Design methodology

7.6.2.1 Pattern analysis

7.6.2.1.1 Histograms, scatter plots, other graphics
7.6.2.1.2 Cluster analysis algorithms
7.6.2.1.3 Mappings
7.6.2.1.4 Exploratory data analysis
7.6.2.1.5 Density estimation procedures

7.6.2.2 Feature evaluation and selection

7.6.2.2.1 Distance measures and error bounds
7.6.2.2.2 Subset selection

7.6.2.3 Classifier design

7.6.2.3.1 Single stage
7.6.2.3.2 Hierarchical classifiers
7.6.2.3.3 Grammatical inference

7.6.2.4 Pattern classification experiments

7.6.2.4.1 Using labelled samples
7.6.2.4.2 Using unlabelled samples

7.6.3 Applications

7.6.3.1 Waveform analysis

7. METHODOLOGIES

                7.6.3.1.1 Biomedical
                7.6.3.1.2 Speech
                7.6.3.1.3 NMR spectroscopy patterns
                7.6.3.1.4 Other

        7.6.3.2 Line drawings

                7.6.3.2.1 Optical character recognition
                7.6.3.2.2 Chemical structure diagrams
                7.6.3.2.3 Blueprints, maps
                7.6.3.2.4 Displayed mathematical expressions
                7.6.3.2.5 Other

        7.6.3.3 Images (*see* 7.5.9)
        7.6.3.4 Others

    7.6.4 Implementations

        7.6.4.1 Hardware

                7.6.4.1.1 Input
                7.6.4.1.2 Processors
                7.6.4.1.3 Displays

        7.6.4.2 Software

                7.6.4.2.1 Packages for data analysis
                7.6.4.2.2 Design of interactive systems

PART III. TAXONOMY TREE
7. METHODOLOGIES
7.7 MODELING AND SIMULATION

7.7 MODELING AND SIMULATION

7.7 Modeling and Simulation (see also 2.4.2.1.1.1 and 4.1.3.3.1)

    7.7.1 Characteristics of simulation models

        7.7.1.1 Type
        7.7.1.2 Dynamic
        7.7.1.3 Mathematical/nonmathematical
        7.7.1.4 Deterministic/nondeterministic

    7.7.2 Discrete event simulation models

        7.7.2.1 Purpose
        7.7.2.2 System characteristics

            7.7.2.2.1 Scope
            7.7.2.2.2 Transient behavior
            7.7.2.2.3 Entities
            7.7.2.2.4 Attributes
            7.7.2.2.5 Variable characteristics

                7.7.2.2.5.1 Random/deterministic
                7.7.2.2.5.2 Single-valued/functional
                7.7.2.2.5.3 Exogenous/endogenous
                7.7.2.2.5.4 Independent/dependent
                7.7.2.2.5.5 Global/local

            7.7.2.2.6 Time advance

        7.7.2.3 Simulation languages

            7.7.2.3.1 World-view

                7.7.2.3.1.1 Transaction-oriented
                7.7.2.3.1.2 Event-oriented
                7.7.2.3.1.3 Process-oriented

            7.7.2.3.2 Language orientation

                7.7.2.3.2.1 Statement-oriented
                7.7.2.3.2.2 Block-oriented
                7.7.2.3.2.3 Other

        7.7.2.4 Applications

            7.7.2.4.1 Business and industry
            7.7.2.4.2 Government and social
            7.7.2.4.3 Engineering and science

    7.7.3 Continuous system simulation models

        7.7.3.1 Purpose
        7.7.3.2 System characteristics

            7.7.3.2.1 Scope
            7.7.3.2.2 Transient behavior
            7.7.3.2.3 Steady state behavior
            7.7.3.2.4 Variable characteristics

                7.7.3.2.4.1 Random/deterministic
                7.7.3.2.4.2 Single-valued/functional
                7.7.3.2.4.3 External/internal
                7.7.3.2.4.4 Independent/dependent
                7.7.3.2.4.5 Stable/unstable

7. METHODOLOGIES

                7.7.3.2.5   Time advance

        7.7.3.3   Simulation languages

                7.7.3.3.1   Type

                        7.7.3.3.1.1   General purpose
                        7.7.3.3.1.2   Hybrid
                        7.7.3.3.1.3   Special purpose

                7.7.3.3.2   Language orientation (see 7.7.2.3.2)

        7.7.3.4   Applications

                7.7.3.4.1   Engineering and science
                7.7.3.4.2   Business and industry
                7.7.3.4.3   Government and social

7.7.4   Modeling considerations

        7.7.4.1   Model validation and verification
        7.7.4.2   Random number generation (see also 5.2.6.1)
        7.7.4.3   Model data reduction and analysis
        7.7.4.4   Model portability and documentation

PART III. TAXONOMY TREE
7. METHODOLOGIES
7.8 SORTING AND SEARCHING

7.8 Sorting and Searching

    7.8.1 Sorting algorithms

        7.8.1.1 Internal sorting

            7.8.1.1.1 Selection

                7.8.1.1.1.1 Straight selection
                7.8.1.1.1.2 Tree selection (Tournament)
                7.8.1.1.1.3 Heapsort
                7.8.1.1.1.4 Other

            7.8.1.1.2 Insertion

                7.8.1.1.2.1 Straight insertion
                7.8.1.1.2.2 Shellsort (Diminishing Increment Sort)
                7.8.1.1.2.3 Linked linear list insertion
                7.8.1.1.2.4 Tree insertion sort

            7.8.1.1.3 Exchange

                7.8.1.1.3.1 Partition exchange (Quicksort)
                7.8.1.1.3.2 Merge exchange (Batcher's parallel sort)
                7.8.1.1.3.3 Other

            7.8.1.1.4 Enumeration (counting)
            7.8.1.1.5 Merging
            7.8.1.1.6 Distribution

                7.8.1.1.6.1 Radix
                7.8.1.1.6.2 Address calculation

        7.8.1.2 External sorting

            7.8.1.2.1 Direct access merge

                7.8.1.2.1.1 String management
                7.8.1.2.1.2 Space management

            7.8.1.2.2 Tape merge

                7.8.1.2.2.1 Balanced
                7.8.1.2.2.2 Unbalanced

            7.8.1.2.3 Distributive

                7.8.1.2.3.1 Record management
                7.8.1.2.3.2 Space management

    7.8.2 Searching algorithms

        7.8.2.1 Sequential comparative

            7.8.2.1.1 Unordered list
            7.8.2.1.2 Ordered list

        7.8.2.2 Tree-based comparative (<u>see also</u> 3.1.2.5)

            7.8.2.2.1 Binary search
            7.8.2.2.2 Fibonacci search

7. METHODOLOGIES

      7.8.2.2.3 Higher order trees

    7.8.2.3 Classification argument

      7.8.2.3.1 Truncated index
      7.8.2.3.2 Frequency ordering

    7.8.2.4 List structure

  7.8.3 Programming considerations

    7.8.3.1 Data structures (<u>see</u> <u>also</u> 3.1)
    7.8.3.2 File structures (<u>see</u> <u>also</u> 3.2.3)
    7.8.3.3 Program structures
    7.8.3.4 Hashing
    7.8.3.5 Buffering

  7.8.4 Hardware organization

    7.8.4.1 Processor
    7.8.4.2 Storage

      7.8.4.2.1 Storage mapping
      7.8.4.2.2 Storage size

    7.8.4.3 I/O subsystem

      7.8.4.3.1 Device characteristics
      7.8.4.3.2 Overlap capabilities

    7.8.4.4 System architecture (<u>see</u> <u>also</u> 1.3)

      7.8.4.4.1 Parallel processors
      7.8.4.4.2 Multiprocessors

  7.8.5 Algorithm analysis

    7.8.5.1 Comparisons
    7.8.5.2 Data movements
    7.8.5.3 Storage required
    7.8.5.4 Underlying theory

PART III. TAXONOMY TREE
7. METHODOLOGIES
7.9 COMPUTER GRAPHICS

7.9 Computer Graphics

    7.9.1 Hardware architecture

        7.9.1.1 Display devices (see also 1.4.2.3)

            7.9.1.1.1 Refresh devices

                7.9.1.1.1.1 Vector displays
                7.9.1.1.1.2 Raster displays

            7.9.1.1.2 Image storage display devices
            7.9.1.1.3 Large screen displays
            7.9.1.1.4 Plotters
            7.9.1.1.5 Microfilm recorders
            7.9.1.1.6 Screen copiers
            7.9.1.1.7 Special-purpose hardware/firmware

                7.9.1.1.7.1 Curve generators
                7.9.1.1.7.2 Font/symbol generators
                7.9.1.1.7.3 Transformatiion hardware
                7.9.1.1.7.4 Clipping divider
                7.9.1.1.7.5 Hidden line/surface eliminator
                7.9.1.1.7.6 Other

        7.9.1.2 Input (interaction) devices

            7.9.1.2.1 Location-specifying devices
            7.9.1.2.2 Picture element picking devices
            7.9.1.2.3 Value-generating devices

    7.9.2 Graphics systems (see also 2.3.1)

        7.9.2.1 Single terminal systems
        7.9.2.2 Multi-terminal systems
        7.9.2.3 Remote graphics
        7.9.2.4 Satellite graphics
        7.9.2.5 Network graphics

    7.9.3 Picture (image) generation (see also 7.5)

        7.9.3.1 Object description

            7.9.3.1.1 Linear segmentation of the display file
            7.9.3.1.2 Segment hierarchy in the display file

        7.9.3.2 Viewing operations

            7.9.3.2.1 Windowing
            7.9 3.2.2 Clipping
            7.9.3.2.3 Projections
            7.9.3.2.4 Viewports
            7.9.3.2.5 Boxing and extents
            7.9.3.2.6 Image transformations

    7.9.4 Graphics utilities

        7.9.4.1 Graphics support software

            7.9.4.1.1 Graphical languages and extensions
            7.9.4.1.2 Graphical subroutine packages

        7.9.4.2 Picture description languages

7. METHODOLOGIES

        7.9.4.3   Typical application system design

                7.9.4.3.1 Plotting packages
                7.9.4.3.2 Character and symbol packages
                7.9.4.3.3 Graphics editors/drafting packages

7.9.5   Object modeling

        7.9.5.1 Object hierarchy
        7.9.5.2 Object transformations
        7.9.5.3 Model to picture transformation

7.9.6   Methodology and techniques

        7.9.6.1 Device independence for portability
        7.9.6.2 Modeling versus viewing
        7.9.6.3 Higher level versus basic ("kernel") graphics systems
        7.9.6.4 Interaction techniques

                7.9.6.4.1 Menu picking
                7.9.6.4.2 Zooming/panning
                7.9.6.4.3 Level of detail/extents
                7.9.6.4.4 Interrupt-driven dialogues
                7.9.6.4.5 Interactive recognizers

        7.9.6.5 Human factors engineering (ergonomics)

                7.9.6.5.1 Workstation organization
                7.9.6.5.2 Operator interface
                7.9.6.5.3 Programmer interface

7.9.7   Three-dimensional graphics and realism

        7.9.7.1 Hidden line/surface elimination
        7.9.7.2 Surface representation
        7.9.7.3 Texture, shading and lighting models
        7.9.7.4 Aliasing
        7.9.7.5 Dithering
        7.9.7.6 Animation

                7.9.7.6.1 Simple motion dynamics
                7.9.7.6.2 Key frame animation
                7.9.7.6.3 Frame at a time
                7.9.7.6.4 Simulators

PART III. TAXONOMY TREE

8. APPLICATIONS/TECHNIQUES [ILLUSTRATIVE]

PART III. TAXONOMY TREE
8. APPLICATIONS/TECHNIQUES
8.1 BUSINESS DATA PROCESSING

8.1 Business Data Processing

8.1.1 Financial systems

- 8.1.1.1 Accounting
- 8.1.1.2 Cash management
- 8.1.1.3 Asset management

8.1.2 Personnel systems

- 8.1.2.1 Skills inventory
- 8.1.2.2 Payroll
- 8.1.2.3 Employment statistics

8.1.3 Marketing systems

- 8.1.3.1 Sales forecasting
- 8.1.3.2 Sales analysis
- 8.1.3.3 Order entry

8.1.4 Distribution

- 8.1.4.1 Vehicle routing
- 8.1.4.2 Load balancing

8.1.5 Manufacturing systems

- 8.1.5.1 Requirements planning
- 8.1.5.2 Production scheduling
- 8.1.5.3 Inventory control
- 8.1.5.4 Quality control

8.1.6 Management information systems

- 8.1.6.1 Budget analyses
- 8.1.6.2 Product profitability analyses
- 8.1.6.3 Cash flow analyses

8.1.7 Miscellaneous

- 8.1.7.1 Purchasing
- 8.1.7.2 Stockholder records

PART III. TAXONOMY TREE
8. APPLICATIONS/TECHNIQUES
8.2 SCIENTIFIC AND ENGINEERING DATA PROCESSING TECHNIQUES

8.2 Scientific and Engineering Data Processing Techniques

- 8.2.1 Data acquisition

    - 8.2.1.1 Passive instrumentation

        - 8.2.1.1.1 Analog devices
        - 8.2.1.1.2 Digital devices

    - 8.2.1.2 Computer-controlled experiments (see also 2.3.4)

        - 8.2.1.2.1 Special-purpose controllers
        - 8.2.1.2.2 General-purpose controllers

- 8.2.2 Analysis of experimental data

    - 8.2.2.1 Data refinement and presentation

        - 8.2.2.1.1 Smoothing and filtering (see also 5.1.6.2.8 and 7.5.3.3-4)
        - 8.2.2.1.2 Graphical presentation (see also 7.9)
        - 8.2.2.1.3 Transformation of domain

    - 8.2.2.2 Error analysis (see also 5.3.1.3)
    - 8.2.2.3 Statistical analysis (see also 5.2.6)

        - 8.2.2.3.1 Analysis of variance
        - 8.2.2.3.2 Multiple correlation
        - 8.2.2.3.3 Stratified sampling
        - 8.2.2.3.4 Other statistical analyses

    - 8.2.2.4 Data reduction

        - 8.2.2.4.1 Least squares curve fitting (see also 5.1.6.2.1 and 5.1.7.1)
        - 8.2.2.4.2 Spline function analysis (see also 5.1.6.1.5 and 5.1.6.2.6)
        - 8.2.2.4.3 Rational approximation (see also 5.1.6.2.5)
        - 8.2.2.4.4 Orthogonal function expansion
        - 8.2.2.4.5 Extraction of parameters

- 8.2.3 Simulation and model building (see also 7.7)

    - 8.2.3.1 Design automation
    - 8.2.3.2 Simulation for prediction
    - 8.2.3.3 Simulation for training

- 8.2.4 Theoretical analysis

    - 8.2.4.1 Arithmetic calculation (see also 5.)

        - 8.2.4.1.1 Solution of equations (see 5.1.1, 5.1.3 and 5.1.4)
        - 8.2.4.1.2 Solution of inequalities (see 5.2.5)
        - 8.2.4.1.3 Monte Carlo simulation (see 5.2.6.2)

    - 8.2.4.2 Algebraic manipulation (see also 7.1)

        - 8.2.4.2.1 Data representation (see 7.1.1)
        - 8.2.4.2.2 Manipulative techniques (see 7.1.2)
        - 8.2.4.2.3 Symbolic languages (see 7.1.3)

PART III. TAXONOMY TREE
8. APPLICATIONS/TECHNIQUES
8.3 COMPUTER-ASSISTED INSTRUCTION

8.3 Computer-Assisted Instruction

8.3.1 Hardware

- 8.3.1.1 Systems
- 8.3.1.2 Communications
- 8.3.1.3 User interface
  - 8.3.1.3.1 Typewriter keyboard
  - 8.3.1.3.2 Graphic display (see also 7.9.1.1)
  - 8.3.1.3.3 Pointers
  - 8.3.1.3.4 Audio input
  - 8.3.1.3.5 Audio output

8.3.2 Software

- 8.3.2.1 Operations
- 8.3.2.2 Development
- 8.3.2.3 Utilities

8.3.3 Applications

- 8.3.3.1 Skills practice
- 8.3.3.2 Diagnostic testing
- 8.3.3.3 Tutorial instruction
- 8.3.3.4 Simulation (see also 7.7)
- 8.3.3.5 Modeling (see also 7.7)
- 8.3.3.6 Problem solving
- 8.3.3.7 Artistic creation
- 8.3.3.8 Materials production

8.3.4 Authoring

- 8.3.4.1 Question-and-answer sequences
- 8.3.4.2 Simulations and models
- 8.3.4.3 Information structures

8.3.5 Evaluation

- 8.3.5.1 Cost
- 8.3.5.2 Effectiveness
- 8.3.5.3 Reliability
- 8.3.5.4 Acceptance

8.3.6 Implications

- 8.3.6.1 Learning
- 8.3.6.2 Social interaction
- 8.3.6.3 Access to information

PART III. TAXONOMY TREE

8. APPLICATIONS/TECHNIQUES

8.4 TEXT (WORD) PROCESSING

8.4 Text (Word) Processing

8.4.1 Text editing

8.4.1.1 Program editors

8.4.1.1.1 Line editors
8.4.1.1.2 Context editors

8.4.1.2 Manuscript editors

8.4.1.2.1 Word processing systems
8.4.1.2.2 Shared logic systems (clusters)
8.4.1.2.3 Time-shared systems

8.4.1.3 Interaction/command specification

8.4.1.3.1 Command language editors
8.4.1.3.2 Screen editors

8.4.1.4 On-line manuscript exploration facilities

8.4.2 Output creation

8.4.2.1 Graphic arts/typography
8.4.2.2 Mark-up systems

8.4.2.2.1 Hyphenation and justification
8.4.2.2.2 Tabular composition
8.4.2.2.3 Mathematical formulas
8.4.2.2.4 Format/typesetting codes

8.4.2.3 Layout and pagination systems

8.4.2.3.1 Noninteractive/automated pagination
8.4.2.3.2 Interactive layout and pagination
8.4.2.3.3 Page makeup terminals

8.4.3 Auxiliary services

8.4.3.1 Routing, queueing and electronic mail
8.4.3.2 Picture handling (graphics)
8.4.3.3 Automatic index generation
8.4.3.4 Spelling verification
8.4.3.5 Journalling and version management

8.4.4 Input

PART III. TAXONOMY TREE
9. COMPUTING MILIEUX

PART III. TAXONOMY TREE
9. COMPUTING MILIEUX
9.1 THE COMPUTER INDUSTRY

9.1 The Computer Industry

    9.1.1 Statistics

        9.1.1.1 Employment
        9.1.1.2 Installation
        9.1.1.3 Financial (see also 9.1.3)

    9.1.2 Markets

        9.1.2.1 Product

            9.1.2.1.1 Hardware (see also 9.1.3.1)
            9.1.2.1.2 Software (see also 9.1.3.2)
            9.1.2.1.3 Computing services (see also 9.1.3.3, 9.1.3.5, 9.1.3.6)
            9.1.2.1.4 Communication services (see also 9.1.3.4)

        9.1.2.2 Geographical

            9.1.2.2.1 Domestic
            9.1.2.2.2 Foreign

    9.1.3 Suppliers

        9.1.3.1 Hardware

            9.1.3.1.1 CPUs
            9.1.3.1.2 Peripheral equipment
            9.1.3.1.3 Terminals
            9.1.3.1.4 Data entry systems
            9.1.3.1.5 Communications
            9.1.3.1.6 Other

        9.1.3.2 Software

            9.1.3.2.1 System software
            9.1.3.2.2 Application software packages
            9.1.3.2.3 Custom software
            9.1.3.2.4 Other

        9.1.3.3 Computing services (see also 9.5.4.1)

            9.1.3.3.1 Service bureaus
            9.1.3.3.2 Time-sharing (see also 2.2.3.1)
            9.1.3.3.3 Data banks
            9.1.3.3.4 Other

        9.1.3.4 Communications services

            9.1.3.4.1 Packet switching (see also 3.4.2.2.2)
            9.1.3.4.2 Common carriers

        9.1.3.5 Professional and management services (see also 9.5.4.2)

            9.1.3.5.1 Consultation
            9.1.3.5.2 Education and training
            9.1.3.5.3 Newsletters and manuals
            9.1.3.5.4 Facilities management
            9.1.3.5.5 Placement and recruiting
            9.1.3.5.6 Maintenance

        9.1.3.6 Leasing companies

PART III. TAXONOMY TREE
9. COMPUTING MILIEUX
9.2 EDUCATION AND COMPUTING

9.2 Education and Computing

    9.2.1 Education about computing

        9.2.1.1 Graduate level

            9.2.1.1.1 Computer science
            9.2.1.1.2 Information systems
            9.2.1.1.3 Computer engineering
            9.2.1.1.4 Service courses

        9.2.1.2 Undergraduate level

            9.2.1.2.1 Computer science
            9.2.1.2.2 Information systems
            9.2.1.2.3 Computer engineering
            9.2.1.2.4 Computational mathematics
            9.2.1.2.5 Minor programs
            9.2.1.2.6 Service courses
            9.2.1.2.7 Computer literacy

        9.2.1.3 Junior and community colleges

            9.2.1.3.1 Programmer trainee programs
            9.2.1.3.2 Data processing programs
            9.2.1.3.3 Service courses
            9.2.1.3.4 Computer literacy

        9.2.1.4 Secondary schools

            9.2.1.4.1 Programming instruction
            9.2.1.4.2 Computer literacy
            9.2.1.4.3 Data processing instruction

        9.2.1.5 Continuing education

            9.2.1.5.1 Seminars and short courses
            9.2.1.5.2 Self-assessment
            9.2.1.5.3 Certification (see also 9.6.4.1)

    9.2.2 Educational uses of computers (see also 8.3)

        9.2.2.1 Higher education
        9.2.2.2 Precollege education
        9.2.2.3 Continuing education

9.3 HISTORY OF COMPUTING

PART III. TAXONOMY TREE
9. COMPUTING MILIEUX

9.3 History of Computing

    9.3.1 Origins (pre-twentieth Century)

        9.3.1.1 Beginnings

            9.3.1.1.1 Abacus
            9.3.1.1.2 Slide rules
            9.3.1.1.3 Napier's bones

        9.3.1.2 Early calculators

            9.3.1.2.1 Pascal's calculator
            9.3.1.2.2 Leibniz' calculator
            9.3.1.2.3 Other early calculators

        9.3.1.3 Charles Babbage

            9.3.1.3.1 The difference engine
            9.3.1.3.2 The analytical engine

        9.3.1.4 Calculators in the nineteenth century

            9.3.1.4.1 Burroughs
            9.3.1.4.2 Odhner
            9.3.1.4.3 Other calculators

    9.3.2 Early developments (until 1950)

        9.3.2.1 Punch card machines

            9.3.2.1.1 Early developments

                9.3.2.1.1.1 Hollerith
                9.3.2.1.1.2 Powers
                9.3.2.1.1.3 Use in census from 1890-1910

            9.3.2.1.2 Later punch card machines

        9.3.2.2 Electromechanical and relay computers

            9.3.2.2.1 Differential analyzers and other analog computers
            9.3.2.2.2 Bell Telephone Laboratories relay computers
            9.3.2.2.3 Zuse Z3
            9.3.2.2.4 Mark I (IBM and Harvard U.)
            9.3.2.2.5 Mark II - Harvard relay computer
            9.3.2.2.6 IBM SSEC

        9.3.2.3 Early electronic computers

            9.3.2.3.1 Vacuum tube flip-flops (Eccles-Jordan)
            9.3.2.3.2 Atanasoff's computer project
            9.3.2.3.3 Colossus
            9.3.2.3.4 The ENIAC
            9.3.2.3.5 The EDVAC
            9.3.2.3.6 Stored program computers

    9.3.3 Recent history

        9.3.3.1 The first generation (vacuum tube computers)

            9.3.3.1.1 Early large-scale computers

9. COMPUTING MILIEUX

    9.3.3.1.2 Medium-scale computers
    9.3.3.1.3 Magnetic core memory computers
    9.3.3.1.4 First generation software systems

      9.3.3.1.4.1 Assemblers
      9.3.3.1.4.2 Compilers
      9.3.3.1.4.3 Other

  9.3.3.2 The second generation

    9.3.3.2.1 Early transistorized computers
    9.3.3.2.2 Major second generation computers
    9.3.3.2.3 Early virtual memory systems

  9.3.3.3 The third generation

    9.3.3.3.1 Hardware (<u>see</u> <u>also</u> 1.)

      9.3.3.3.1.1 Integrated circuits
      9.3.3.3.1.2 IBM 360 and 370 series
      9.3.3.3.1.3 Other major computer systems
      9.3.3.3.1.4 Large-scale integration
      9.3.3.3.1.5 Minicomputers
      9.3.3.3.1.6 Microprocessors

    9.3.3.3.2 Software (<u>see</u> <u>also</u> 4.)

      9.3.3.3.2.1 Operating systems
      9.3.3.3.2.2 Time-sharing systems
      9.3.3.3.2.3 Other software systems

    9.3.3.3.3 Systems

      9.3.3.3.3.1 Communications systems
      9.3.3.3.3.2 Array processors
      9.3.3.3.3.3 Vector processors

PART III. TAXONOMY TREE
9. COMPUTING MILIEUX
9.4 LEGAL ASPECTS OF COMPUTING

9.4 Legal Aspects of Computing

9.4.1 Software protection mechanisms

- 9.4.1.1 Copyright
- 9.4.1.2 Patent
- 9.4.1.3 Trade secret
- 9.4.1.4 Registration
- 9.4.1.5 Contract

9.4.2 Privacy

- 9.4.2.1 Data banks
- 9.4.2.2 Transmission systems
- 9.4.2.3 Transborder data flow (see also 9.7.6.1)

9.4.3 Security (see also 9.5.5.1)

- 9.4.3.1 Civil sanctions
- 9.4.3.2 Criminal sanctions

9.4.4 Taxation

- 9.4.4.1 Hardware
- 9.4.4.2 Software
- 9.4.4.3 Data processing services

9.4.5 Contracts

- 9.4.5.1 Hardware acquisition
  - 9.4.5.1.1 Sales contracts
  - 9.4.5.1.2 Lease contracts
- 9.4.5.2 Software acquisition
  - 9.4.5.2.1 Sales contracts
  - 9.4.5.2.2 Lease contracts
- 9.4.5.3 Maintenance
- 9.4.5.4 Services
  - 9.4.5.4.1 Processing
  - 9.4.5.4.2 Facilities management
  - 9.4.5.4.3 Licenses
  - 9.4.5.4.4 Escrow of software in source form
- 9.4.5.5 Other

9.4.6 Computer-communications interface

- 9.4.6.1 FCC inquiries
- 9.4.6.2 Datapaths
  - 9.4.6.2.1 Satellites
  - 9.4.6.2.2 Specialized common carriers

9.4.7 Antitrust

- 9.4.7.1 Government litigation
- 9.4.7.2 Private suits

9.4.8 Other legislation/regulation

                9.4.8.1   Import/export
                9.4.8.2   Procurement
                9.4.8.3   Banking/EFT (see also 2.3.4.6.3)

                          9.4.8.3.1   Automated teller machines
                          9.4.8.3.2   Customer bank communication terminals

                9.4.8.4   Personnel

                          9.4.8.4.1   Labor relations
                          9.4.8.4.2   Licensing

                9.4.8.5   Consumer credit

        9.4.9   The computer in litigation

                9.4.9.1   Admissibility of computerized records
                9.4.9.2   Discovery of computerized records

PART III. TAXONOMY TREE
9. COMPUTING MILIEUX
9.5 MANAGEMENT OF COMPUTING

9.5 Management of Computing

9.5.1 Management support

9.5.1.1 Long range planning

- 9.5.1.1.1 Capacity
- 9.5.1.1.2 Applications
- 9.5.1.1.3 Personnel
- 9.5.1.1.4 Financial
- 9.5.1.1.5 Facilities
- 9.5.1.1.6 Communications network

9.5.1.2 Financial management

- 9.5.1.2.1 Budgets
- 9.5.1.2.2 Cost analyses
- 9.5.1.2.3 Charges for services

9.5.1.3 Installation standards

- 9.5.1.3.1 Hardware
- 9.5.1.3.2 Software
- 9.5.1.3.3 Interfaces
- 9.5.1.3.4 Documentation (see also 4.1.3.4, 9.5.2.2.5 and 9.5.5.4.4)
- 9.5.1.3.5 Communications

9.5.1.4 Auditing (see also 9.7.1.4)

- 9.5.1.4.1 Requirements and methodologies
- 9.5.1.4.2 External auditors

9.5.2 Application system development

9.5.2.1 Project initiation

- 9.5.2.1.1 Requirements definition
- 9.5.2.1.2 Functional specifications
- 9.5.2.1.3 Feasibility studies

9.5.2.2 System development

- 9.5.2.2.1 Design
- 9.5.2.2.2 Programming
- 9.5.2.2.3 Module testing
- 9.5.2.2.4 System testing
- 9.5.2.2.5 Documentation (see also 4.1.3.4, 9.5.1.3.4 and 9.5.5.4.4)
- 9.5.2.2.6 Acceptance testing

9.5.2.3 Implementation

- 9.5.2.3.1 Conversion
- 9.5.2.3.2 Training
- 9.5.2.3.3 Post-implementation audit

9.5.2.4 Application system maintenance (see also 9.5.5.4.3)

- 9.5.2.4.1 Error detection, reporting and correction
- 9.5.2.4.2 System enhancement
- 9.5.2.4.3 Change control

9. COMPUTING MILIEUX

        9.5.2.5   Project management

                9.5.2.5.1 Schedule
                9.5.2.5.2 Manpower
                9.5.2.5.3 Cost

9.5.3   Advanced development

        9.5.3.1   Hardware

                9.5.3.1.1 Evaluation
                9.5.3.1.2 Role of minicomputers and microprocessors
                9.5.3.1.3 Intelligent terminals
                9.5.3.1.4 Interfaces

        9.5.3.2   Software

                9.5.3.2.1 Operating systems
                9.5.3.2.2 Programming methodology
                9.5.3.2.3 Communications network control

        9.5.3.3   User systems

                9.5.3.3.1 Database management (see also 7.4)
                9.5.3.3.2 Word processing (see also 8.4)
                9.5.3.3.3 Office automation

        9.5.3.4   Delivery of services

                9.5.3.4.1 Networks
                9.5.3.4.2 Distributed processing
                9.5.3.4.3 Interactive, real time and batch

9.5.4   User services

        9.5.4.1   Computing services (see also 9.1.3.3)

                9.5.4.1.1 Local
                9.5.4.1.2 Remote

        9.5.4.2   Professional and management services (see also 9.1.3.5)

                9.5.4.2.1 Consultation
                9.5.4.2.2 Design and programming
                9.5.4.2.3 Education and training

9.5.5   Operations

        9.5.5.1   Security (see also 7.4.5.9 and 9.4.3)

                9.5.5.1.1 Physical
                9.5.5.1.2 Data
                9.5.5.1.3 Program
                9.5.5.1.4 System

        9.5.5.2   Computer operations (see also 4.2.3.1)

                9.5.5.2.1 Console
                9.5.5.2.2 Peripheral
                9.5.5.2.3 Remote

        9.5.5.3   Support operations

9.5 MANAGEMENT OF COMPUTING

          9.5.5.3.1 Data entry
          9.5.5.3.2 Scheduling
          9.5.5.3.3 Dispatching
          9.5.5.3.4 Library
          9.5.5.3.5 Quality control

      9.5.5.4 Maintenance

          9.5.5.4.1 Hardware
          9.5.5.4.2 System software
          9.5.5.4.3 Application software (see also 9.5.2.4)
          9.5.5.4.4 Documentation (see also 4.1.3.4, 9.5.1.3.4 and 9.5.2.2.5)

      9.5.5.5 Capacity management

          9.5.5.5.1 Performance measurement (see also 2.4)
          9.5.5.5.2 Configuration management
          9.5.5.5.3 Reliability/availability

9.5.6 Organization (see also 9.7.3)

      9.5.6.1 Philosophy

          9.5.6.1.1 Centralization/decentralization issues (see also 7.4.6.3)
          9.5.6.1.2 Cost center/profit center issues

      9.5.6.2 External relationships

          9.5.6.2.1 Reporting relationships
          9.5.6.2.2 Use of advisory committees
          9.5.6.2.3 User liaison

      9.5.6.3 Internal organization

          9.5.6.3.1 Structure
          9.5.6.3.2 Job descriptions (see also 9.6.1)

      9.5.6.4 Personnel management

          9.5.6.4.1 Personnel evaluation
          9.5.6.4.2 Professional development
          9.5.6.4.3 Career path planning
          9.5.6.4.4 Salary administration

9.6 THE COMPUTING PROFESSION

PART III. TAXONOMY TREE
9. COMPUTING MILIEUX

9.6　The Computing Profession

　　9.6.1　Occupational titles

　　　　9.6.1.1　Computer scientist
　　　　9.6.1.2　Computer engineer
　　　　9.6.1.3　Systems analyst
　　　　9.6.1.4　Programmer
　　　　9.6.1.5　Computer operator
　　　　9.6.1.6　Other

　　9.6.2　Organizations

　　　　9.6.2.1　Technical societies

　　　　　　9.6.2.1.1　AFIPS
　　　　　　9.6.2.1.2　ACM
　　　　　　9.6.2.1.3　DPMA
　　　　　　9.6.2.1.4　IEEE Computer Society
　　　　　　9.6.2.1.5　BCS
　　　　　　9.6.2.1.6　IFIP
　　　　　　9.6.2.1.7　Other

　　　　9.6.2.2　User groups

　　　　　　9.6.2.2.1　SHARE
　　　　　　9.6.2.2.2　GUIDE
　　　　　　9.6.2.2.3　USE
　　　　　　9.6.2.2.4　VIM
　　　　　　9.6.2.2.5　DECUS
　　　　　　9.6.2.2.6　Other

　　　　9.6.2.3　Trade associations

　　　　　　9.6.2.3.1　ADAPSO
　　　　　　9.6.2.3.2　CBEMA
　　　　　　9.6.2.3.3　Other

　　　　9.6.2.4　ICCP

　　9.6.3　Professional ethics

　　　　9.6.3.1　Canons
　　　　9.6.3.2　Considerations
　　　　9.6.3.3　Rules

　　9.6.4　Testing programs

　　　　9.6.4.1　Certification *see* *also* 9.6.2.4)

　　　　　　9.6.4.1.1　CCP
　　　　　　9.6.4.1.2　CDP

　　　　9.6.4.2　Self-assessment

　　9.6.5　Licensing

PART III. TAXONOMY TREE
9. COMPUTING MILIEUX

9.7 SOCIAL ISSUES AND IMPACTS OF COMPUTING

9.7 Social Issues and Impacts of Computing

9.7.1 Theory and methods

- 9.7.1.1 Theoretical perspectives
- 9.7.1.2 Historical analyses
- 9.7.1.3 Empirical methods
- 9.7.1.4 Futures and forecasting
- 9.7.1.5 Models for evaluating social effects
  - 9.7.1.5.1 Utilitarian criteria
  - 9.7.1.5.2 Non-utilitarian criteria

9.7.2 Growth and development of computing

- 9.7.2.1 Diffusion of computing technology
- 9.7.2.2 Transfer of computing technology
- 9.7.2.3 The computing world
  - 9.7.2.3.1 Social organization
  - 9.7.2.3.2 Ideologies
  - 9.7.2.3.3 Expert-lay interaction

9.7.3 Organizations (see also 9.5.6)

- 9.7.3.1 Workplace
  - 9.7.3.1.1 Work styles and job characteristics
  - 9.7.3.1.2 Employment and skills
  - 9.7.3.1.3 Productivity
  - 9.7.3.1.4 Work organization
- 9.7.3.2 Organizational structure
  - 9.7.3.2.1 Organizational power
  - 9.7.3.2.2 Patterns of control
- 9.7.3.3 Decision making
  - 9.7.3.3.1 Policy making
  - 9.7.3.3.2 Decision style
- 9.7.3.4 Production of goods and services
- 9.7.3.5 Dependency upon computer-based services
- 9.7.3.6 Reliability of computer-based systems (see also 9.5.1.4)

9.7.4 Public policy issues and societal impacts

- 9.7.4.1 Public attitudes and perceptions of computing
- 9.7.4.2 Employment and labor markets
- 9.7.4.3 Privacy and surveillance
- 9.7.4.4 Computing and electoral/legislative activity
  - 9.7.4.4.1 Voting
  - 9.7.4.4.2 Political campaign practices
- 9.7.4.5 Machine intelligence and human identity

9.7.5 Computing in industrial societies

- 9.7.5.1 Social organization

9. COMPUTING MILIEUX

      9.7.5.1.1 Everyday life
      9.7.5.1.2 Inter-institutional and sectoral arrangements

    9.7.5.2 Publics and computer-using organizations
    9.7.5.3 Service provision
    9.7.5.4 Accountability of computing
    9.7.5.5 Industrial growth
    9.7.5.6 Computing and cultural values

  9.7.6 Cross-national impacts

    9.7.6.1 Transborder data flow (<u>see also</u> 9.4.2.3)
    9.7.6.2 Computing in less developed countries
    9.7.6.3 Multinational organizations and services
    9.7.6.4 Technology transfer
    9.7.6.5 National sovereignty and balance of trade

  9.7.7 Humanistic studies of computing

    9.7.7.1 Philosophical inquiries

      9.7.7.1.1 Ethical issues of computer use
      9.7.7.1.2 Philosophical anthropology
      9.7.7.1.3 Philosophy of technology

    9.7.7.2 Computing in literature
    9.7.7.3 Computing in the arts

IV. ANNOTATED TAXONOMY TREE

PART IV. ANNOTATED TAXONOMY TREE
1. HARDWARE

1. Hardware

[All the elements, devices, components, subsystems as well as their organizations and functions that are used in computer systems.]

1.1 Types of Computers

1.1.1 Digital computer

[A computer which processes data in discrete form or by modeling the quantities in computation by analogous discrete quantities.]

1.1.1.1 Microcomputer

[A digital computer whose central processing unit (see 1.2.1) is implemented on a single integrated circuit chip with other chips for additional memory and input/output functions; smallest physical size and computing power among all types of digital computers; as of 1979, typically has an 8-bit word length (though 4-bit and 16-bit word length machines are also common), and a speed ranging from 5 to 50 thousand instructions per second.]

1.1.1.2 Minicomputer

[A digital computer whose central processing unit is implemented by a number of circuit boards; larger instruction set and computing power than that of a microcomputer and generally has one or more mass memory devices and a limited amount of input/output devices; as of 1979, typically has a 16-bit word length (though 7-bit, 12-bit, 18-bit, 24-bit and 32-bit word length machines are also common), and speed ranging from 0.05 to 0.5 million instructions per second.]

1.1.1.3 Medium-scale computer

[A digital computer which generally has larger computing power and longer word length than that of a minicomputer, and containing a variety of input/output devices; as of 1979, typically has a 24-bit word length (though 16-bit and 36-bit word length machines are also common), and speed ranging from 0.5 to 5 million instructions per second.]

1.1.1.4 Large-scale computer

[A digital computer which is generally the largest and fastest available with standard features; normally has a large amount of mass memory and other peripheral devices; as of 1979, typically has a word length of 48 or 64 bits (though some may have as small as 32-bit word length) and speed of 5 to 50 million instructions per second.]

1.1.1.5 Supercomputer

[A digital computer which is generally the largest and fastest among all types of computers; word length may be as large as 128 bits and, as of 1979, may have

1. HARDWARE

speed up to 500 million instructions per second.]

1.1.2 Analog computer

[A system which establishes prescribed relations between continuous variables and physical quantities; a typical use is the modeling of a dynamic system such as an automobile suspension system.]

1.1.3 Hybrid computer

[A computer which includes both analog and digital components; typically digital components are used for control and for auxiliary computation.]

1.1.3.1 Basic patchable hybrid computer

[An analog computer with minimal digital components used to make logical decisions which change program parameters, simulate actual events, and provide limited optimization capability.]

1.1.3.2 Patchable clocked system

[A typical system includes a digital clock and several registers; such a system permits program cycling and counting and allows small adaptive control studies.]

1.1.3.3 Mini-hybrid computer

[A system consisting of an analog computer with an interface to a small digital computer.]

1.1.3.4 Full hybrid computer

[An analog computer that is usually quite large and includes a full complement of patchable digital logic.]

1.2 Digital Computer Subsystems

1.2.1 Central processing unit (CPU)

[The unit of a digital computer that includes the circuits controlling the interpretation and execution of instructions.]

1.2.1.1 Arithmetic-logic unit

[Unit in which arithmetic, logic and related operations are performed.]

1.2.1.2 Control unit

[The part of a digital computer that selects a coded instruction, interprets it, develops proper signals or signal sequences, and applies them to various parts of the computer in order to achieve execution of the instruction.]

1.2.1.2.1 Hardwired

[Where an instruction initiates a sequence of elementary micro-operations fixed at the time of manufacture.]

1. HARDWARE

1.2.1.2.2 Microprogrammed

[Where an instruction initiates a sequence of elementary micro-instructions, called a microprogram, which are stored in a storage unit and may be altered.]

1.2.2 Memory

[A device which receives data, stores data and, when requested, returns data.]

1.2.2.1 Types

1.2.2.1.1 Cache memory

[A small high-speed storage unit which, combined with a large, slower store, forms a storage hierarchy to achieve high performance.]

1.2.2.1.2 Main memory

[A memory device whose cells can be addressed by a computer program and from which instructions and data can be loaded directly into registers for execution.]

1.2.2.1.3 Mass memory

[A relatively large-volume, slow-speed storage such as a magnetic drum, disk or tape.]

1.2.2.1.4 Buffer memory

[A memory device used to interface the communication of two devices that have different rates of data flow, different data formats, and different time of occurrence of events.]

1.2.2.2 Memory access modes

[How a data record is stored into or retrieved from a memory.]

1.2.2.2.1 Coordinate-addressed access

[A data record is stored into or retrieved from memory according to its coordinate, or location, address.]

1.2.2.2.1.1 Random access

[The storage and retrieval time is independent of the location of the data record.]

1.2.2.2.1.2 Sequential access

[The access time depends on how far away the desired data

1. HARDWARE

> record is located at the beginning of fetching.]

1.2.2.2.1.3 Multidimensional access

[Access can be in groups of data of different sizes such as a bit slice, a number of bytes, or a number of words.]

1.2.2.2.2 Content-addressed access

[A data record is stored into or retrieved from memory (called content addressable or associative memory) according to its content rather than location or address.]

1.2.3 Input/Output (I/O)

[Input/Output refers to the hardware associated with the transfer of information, data, status and control between a processor (CPU plus main memory) and its environment.]

1.2.3.1 Communication path

[The route by which I/O information enters and leaves the processor.]

1.2.3.1.1 Register-based I/O

[I/O access to the processor is directed to one or more registers of the CPU; thus I/O information is directly programmer accessible.]

1.2.3.1.2 Memory-based I/O

[I/O access to the processor is directly to memory; thus the CPU obtains the I/O information by means of memory references.]

1.2.3.1.2.1 CPU resident controller

[Where the hardware directing the flow of information to memory is an integral part of the CPU.]

1.2.3.1.2.2 I/O processor (I/O channel)

[Where the hardware directing the flow of information to memory is external to the CPU; thus, the CPU and I/O processor compete for memory access.]

1.2.3.2 Signaling mechanism

[The means by which control of I/O transfers is effected.]

1.2.3.2.1 Programmed I/O

1. HARDWARE

[Aspects of the I/O transfer are under the programmer's jurisdiction including verifying in software that an external device is ready for the transfer.]

1.2.3.2.2 Interrupt driven I/O

[The I/O device signals the CPU of an event, such as completion of a data transfer, which results in an interruption of whatever processing is currently in progress.]

1.2.3.3 Input/Output addressing

[The manner by which a particular I/O path is accessed.]

1.2.3.3.1 Separate I/O space

[I/O access is physically distinct from memory access.]

1.2.3.3.2 Memory mapped

[The same physical mechanism used to access a memory location as is used to access an I/O location.]

1.3 Digital Computer Architecture

[The organization of a digital computer usually with emphasis on the processing units.]

1.3.1 Sequential processor

[A computer system in which the processor executes instructions sequentially.]

1.3.2 Parallel processor

[A computer system consisting of multiple processing units and memory modules to accomplish parallel processing.]

1.3.2.1 Homogeneous parallel processors

[A parallel processor in which all processing units are identical.]

1.3.2.1.1 Associative processor

[A homogeneous parallel processor capable of simultaneous searching of all data items and storing/retrieving data items based on part or all of their content.]

1.3.2.1.2 Ensemble processor

[A homogeneous parallel processor in which there is no direct communication among processing units.]

1. HARDWARE

1.3.2.1.3 Array processor

[A homogeneous parallel processor having one or more circuits for communication among processing units.]

1.3.2.1.4 Multiprocessor

[A homogeneous (but sometimes heterogeneous; see 1.3.2.2) parallel processor in which processing units are under the control of a single integrated operating system.]

1.3.2.1.5 Vector processor

[A homogeneous parallel processor capable of processing a vector of data by its processing units functioning in an assembly-line fashion.]

1.3.2.2 Heterogeneous parallel processors

[A parallel processor in which processing units are not all identical.]

1.3.2.2.1 Pipelined machine

[A heterogeneous parallel processor consisting of a number of dedicated processing units functioning in an assembly-line fashion.]

1.3.2.2.2 Data flow machine

[A heterogeneous parallel processor capable of performing data-driven parallel execution of programs represented in data flow form.]

1.3.2.2.3 Distributed processors

[A heterogeneous parallel processor consisting of a number of geographically separated processing units with cooperating control and partitioned data base; (see also 2.1.2.1).]

1.3.2.2.4 Computer network

[An interconnection of an assembly of geographically separated computer systems.]

1.4 Input/Output Devices

[Computer hardware by which data and/or instructions are entered into a computer or by which intermediate or final results are recorded for immediate or future use.]

1.4.1 Data entry/retrieval devices

[A device that is used to convert information into a machine recognizable form and used to convert the results of the

1. HARDWARE

computations into a form usable and manipulable by humans.]

1.4.1.1 Card reader/punch

[A unit which senses and translates data on a punched card into internal computer form or punches onto a card data from a computer after translation.]

1.4.1.2 Paper tape reader/punch

[A unit capable of sensing and translating data on punched paper (or plastic) tape into internal computer form or of punching data onto punched tape which has been translated from internal computer form.]

1.4.1.3 Printer

[A unit for converting internal character representations into symbols on a printed paper.]

1.4.1.3.1 Line printer

[All characters in one line are printed simultaneously.]

1.4.1.3.2 Character printer

[Each character is printed at a separate instant of time.]

1.4.1.3.3 Laser printer

[Characters are burned in by a laser beam.]

1.4.1.4 Plotter

[A unit which inscribes data graphically on a sheet of paper as a function of one or more variables (see also 7.9.1.1.4).]

1.4.1.5 Optical scanner

[A unit which employs optics to convert printed or written data into machine recognizable form.]

1.4.1.6 Microfilm recorder

[A unit that converts digital representation of data to readable or graphical form on film (see also 7.9.1.1.5).]

1.4.1.7 Speech recognizer

[A unit that converts input in speech form into machine recognizable form.]

1.4.2 Terminals

[Devices which directly communicate with the computer system or subsystem in an on-line, interactive fashion.]

1.4.2.1 Keyboard/printer terminal

1. HARDWARE

[A device much like a typewriter which allows an operator to enter data via a keyboard and on which the computer responds by typing messages and results.]

1.4.2.1.1 Teletypes and impact devices

[The printing mechanism is similar to that on a typewriter.]

1.4.2.1.2 Matrix printing devices

[Each character printed is made up of a pattern of dots from a matrix, often 5x7.]

1.4.2.1.3 Non-impact devices

[Techniques such as xerographic, thermal or electrostatic are used to print characters without any impact of a key or matrix on the paper.]

1.4.2.2 Cathode ray tube (CRT) terminal

[A device which allows an operator to enter data via a keyboard and on which computer responses are displayed on a TV-like monitor.]

1.4.2.3 Graphic terminal

[A device which pictorially displays data and allows the operator to enter data via a keyboard, lightpen, function keys or other devices (see also 7.9.1.1.1-3).]

1.4.2.4 Intelligent terminal

[A device which has some data processing capability in addition to its normal input or output function.]

1.4.3 Peripheral devices

[Devices that are used to retain both intermediate results and final computations for immediate and future processing in a computer system.]

1.4.3.1 Magnetic tape device

[Units which transfer binary data to and from a plastic or metal tape which is impregnated or coated with a magnetic material.]

1.4.3.1.1 Cassette

[The tape is packaged in a cassette such as is used in tape recorders.]

1.4.3.1.2 Cartridge

[The tape is packaged in a small enclosed cartridge.]

1.4.3.1.3 Reel

1. HARDWARE

[The tape is on a reel which is not further enclosed.]

1.4.3.2 Magnetic disk and drum devices

[A unit which transfers data to and from a rotating disk or cylinder.]

1.4.3.2.1 Floppy disk

[The disk consists of a flexible magnetic material on which reading-recording heads access data along concentric circular paths.]

1.4.3.2.2 Cartridge disk

[The disk is a single solid platter.]

1.4.3.2.3 Disk pack

[The storage medium consists of two or more solid platters mounted in a single unit.]

1.4.3.2.4 Drum

[A cylinder of magnetic material in which the reading-recording heads access data along parallel paths around the circumference.]

1.4.3.3 Magnetic card devices

[A unit which transfers data to and from a card impregnated or coated with a magnetic material.]

1.4.3.4 Display devices

[Units on which data is exhibited to a user.]

1.4.3.4.1 CRT

[The screen consists of a glass coated with a phosphor that glows on impact by a beam of electrons (see 1.4.2.2).]

1.4.3.4.2 TV

[The display is a television monitor converted to display information carried by digital signals.]

1.4.3.4.3 Plasma

[The screen contains an illuminant gas that glows on application of current conducted to specific points by transparent conductors.]

1.4.4 I/O processor

[A processor which handles the input/output for another or host computer, particularly to achieve high speed and/or high capacity.]

- 1.4.4.1 Front-end processor

    [A device which services terminals and/or peripherals.]

- 1.4.4.2 Remote concentrator

    [A device which handles many low-speed terminals and devices (usually remotely located) and transmits their data over a high-speed communications line.]

- 1.4.4.3 Message-switching processor

    [A unit which receives messages or data from terminals or other message-switching processors and routes the data to the computer or other processors as necessary.]

1.4.5 Modems

[Devices which convert digital signals to and from signals appropriate for transmission across common carrier communication facilities.]

- 1.4.5.1 Direct coupler

    [Digital signals are converted directly to signals compatible with the communications system.]

- 1.4.5.2 Acoustic coupler

    [Digital signals are linked to the communications system, which is typically a voice system, by sound waves.]

1.5 Computer Circuitry

1.5.1 Digital circuitry (logic circuitry)

[Units which convert input logic values to output logic values.]

- 1.5.1.1 Combinational circuits

    [The output logic levels depend on present input logic values only.]

    - 1.5.1.1.1 Contact circuit

        [Consists of relays and/or switches whose positions are controlled by input signals.]

    - 1.5.1.1.2 Symmetric circuit

        [Consists of logic units where the outputs are independent of the order in which inputs are applied.]

    - 1.5.1.1.3 Threshold circuit

        [Consists of logic units where the output depends on the relative value of a weighted sum of the inputs and some specified

1. HARDWARE

threshold.]

1.5.1.2 Sequential circuits

[The output logic values depend on past and perhaps present logic values.]

1.5.1.2.1 Mealy model

[The output depends on past and present inputs.]

1.5.1.2.2 Moore model

[The output depends on past inputs only.]

1.5.1.3 Iterative array

[A combinational or sequential circuit composed of a cascade of identical subcircuits.]

1.5.1.4 Special logic circuits

[A digital circuit commercially available in a single integrated circuit package.]

1.5.1.4.1 Flip-flop

[A Moore model sequential circuit used to store a single bit with one independent output, one or more inputs, and a clock input which determines when output values change.]

1.5.1.4.2 Shift register

[A cascade of flip-flops which stores in succession the logic values that appear at the input.]

1.5.1.4.3 Counter

[A connection of flip-flops and perhaps combinational circuits in which the output logic value, when viewed as a nonnegative integer, is incremented or decremented each time the clock input is activated.]

1.5.1.4.4 Comparator

[A combinational circuit that compares two numbers and produces an output signifying that one is larger, equal to, or smaller than the other.]

1.5.1.4.5 Adder

[A digital circuit in which the output represents the sum of two input numbers.]

1.5.1.4.6 Multiplier

[A digital circuit in which the output

represents the product of two input numbers.]

1.5.1.4.7 Universal logic circuit

[A logic circuit which can be made to realize any combinational logic function on a set of inputs by an appropriate interconnection of leads.]

1.5.1.4.8 Multiplexor

[A combinational circuit in which the single output is connected to one of the primary inputs depending on the values applied to secondary inputs.]

1.5.1.4.9 Decoder

[A combinational circuit in which one and only one output is activated for each combination of input values.]

1.5.2 Analog circuitry

[Units in which inputs and outputs are continuously varying values.]

1.5.2.1 Operational amplifier

[The output values are some constant factor times the input value or perhaps the sum of several input values.]

1.5.2.2 Integrator

[The output is the integral of the input over some specified period of time.]

1.5.2.3 Differentiator

[The output is the derivative with respect to time of the input value.]

1.5.3 Hybrid circuitry

[Units in which inputs and outputs are a combination of continuously varying and discrete values.]

1.5.3.1 Analog-to-digital converter (A-to-D converter)

[The input is a continuously varying value and the output is a set of logic values which, when viewed as a number, approximates the input value.]

1.5.3.2 Digital-to-analog converter (D-to-A converter)

[The output takes on a value determined by the number represented by the set of digital input values.]

1.5.3.3 Schmitt trigger

[The output is a logic value determined by whether the

continuously varying input is above or below a threshold.]

1.6 Computer Elements

[Circuits or small devices which perform some specific, elementary data processing function.]

1.6.1 Semiconductor elements

[A computer element using semiconductor material which displays different electrical resistances in opposite directions.]

1.6.1.1 Bipolar

[The physical process involves both majority and minority carriers; it is called N-type if the predominant, or majority, carriers are electrons and the minority carriers are holes; it is P-type otherwise.]

1.6.1.2 Field-effect

[The physical process involves only majority carriers; sometimes called a unipolar element.]

1.6.2 Magnetic elements

[Devices which utilize the property of magnetic hysteresis for their operations.]

1.6.2.1 Magnetic core

[A magnetic element which has one or more holes through which drive and sense lines are threaded.]

1.6.2.1.1 Single hole

[A toroid-shaped magnetic core which is used mainly as a storage element.]

1.6.2.1.2 Multiaperture

[A magnetic core which has two or more holes through which drive and sense lines are threaded; it is used for both storage and logic functions.]

1.6.2.2 Magnetic thin film

[A thin layer of magnetic material deposited on a substrate; comparable to a magnetic core but is not as widely used.]

1.6.2.2.1 Plated wire

[A layer of magnetic material which is usually electrochemically deposited on a copper wire.]

1.6.2.2.2 Flat film

1. HARDWARE

[One or more layers of magnetic material which are deposited on a flat substrate.]

1.6.2.3 Magnetic bubble device

[A device containing magnetic bubbles which are tiny cylindrical regions of reverse magnetization; specially shaped magnetic structures perform storage or logic functions by controlling the mobility of the domains.]

1.6.3 Charge-coupled device

[A device with a number of closely spaced capacitors on a semiconductor chip.]

1.6.4 Cryogenic element

[A circuit or small device which utilizes the superconductive properties of certain materials near absolute zero temperatures.]

1.6.5 Optical element

[A component part of a system which depends on the properties and phenomena of light.]

1.6.6 Mechanical element

[A component part of a system which depends on the properties and phenomena of force on bodies and with motion; used primarily in early telephone switching systems and machine controllers.]

1.6.7 Hydraulic element

[A component part of a system which functions on the properties and phenomena of liquid in motion.]

1.7 Computer Hardware Reliability

[The extent to which computer hardware functions according to its design.]

1.7.1 Measures

[Quantities which can meaningfully measure computer hardware reliability, such as mean-time-to-failure and mean-time-between-failures.]

1.7.2 Fault classification

1.7.2.1 Number of faults

[The number of faults which may occur in a particular (usually very small) interval.]

1.7.2.1.1 Single fault

[Only a single fault may occur in an interval; this is the so-called single fault assumption which is used in many test generation approaches.]

1.7.2.1.2 Multiple faults

[Any number of faults may occur in an interval.]

1.7.2.2 Types of faults

1.7.2.2.1 Permanent fault

[A fault which continues to occur under the same input; sometimes called a hard fault, solid fault or reproducible fault.]

1.7.2.2.1.1 Stuck-at fault

[A fault whose effect always makes a certain line or lines stuck-at the logic value 0 or 1.]

1.7.2.2.1.2 Shorted fault

[A fault which is caused by an undesired shorted circuit.]

1.7.2.2.2 Transient fault

[A fault which does not always occur under the same input; sometimes called a soft or indeterminate fault.]

1.7.3 Methodologies

[Methodologies to improve computer hardware reliability.]

1.7.3.1 Element improvement

[Methodologies that improve the reliability of computer elements, most involving the improvement of processes in producing computer elements.]

1.7.3.2 Component improvement

[Methodologies that improve the reliability of components.]

1.7.3.2.1 Static redundancies

[Methodologies which use built-in redundant parts to mask out faulty components and do not involve reconfiguration, such as triple modular redundancy.]

1.7.3.2.2 Dynamic redundancies

[Methodologies which use built-in redundant parts to automatically replace faulty components, such as duplication.]

1.7.3.2.3 Hybrid redundancies

[Methodologies which use built-in redundant parts to mask out faulty components and

1. HARDWARE

then switch in spare redundant parts to restore the static redundancy.]

1.7.4 Test generation approaches

[Methods to generate test cases for detecting and locating faults.]

1.7.5 Fault simulation

[A method which simulates the behavior of the computer hardware in the presence of faults, and is normally used as an aid for test generation and computer-aided design of the computer hardware.]

PART IV. ANNOTATED TAXONOMY TREE
2. COMPUTER SYSTEMS

2. Computer Systems

[Systems consisting of more than a single computer.]

2.1 Structure-based Systems

[Systems which are general purpose in the sense that they are intended for a wide variety of applications but in which the system structure or organization is the distinguishing characteristic; cf. 2.3.]

2.1.1 Hybrid systems

[Consisting of both analog and digital components.]

2.1.2 Multiple processor systems

[Digital systems with several central processing units (CPUs).]

2.1.2.1 Distributed systems

[Consisting of several, generally equally important but geographically separated nodes, each capable of doing part of the information processing; see also 1.3.2.2.3.]

2.1.2.2 Hierarchical systems

[Systems with two or more levels of processors, distinguished primarily by control of resources or scheduling, or by processing function.]

2.1.2.2.1 Tightly-coupled systems

[Where most processes involve several processors in close communication, possibly sharing primary storage.]

2.1.2.2.2 Loosely-coupled systems

[With separate processors carrying out independent functions and only communicating with regard to transfer of large tasks between them.]

2.1.2.3 Shared memory systems

[Where several processors share a common primary memory; usually processors are similar in function and capability.]

2.1.2.4 Master-slave systems

[Where there are several processors, usually each with independent primary storage, with one processor allocating tasks to the others.]

2.1.3 Parallel systems

[Systems involving large-scale components acting in parallel.]

2.1.3.1 Pipeline processors

[Different stages of the same process (e.g.

multiplication) may be carried out simultaneously on different data.]

- 2.1.3.2 Array of processors

    [Multiple processing units acting under the coordination of a central control.]

- 2.1.3.3 Multiple functional units

    [Multiple processing units acting asynchronously or as directed by the data.]

2.2 Access-based Systems

[Systems which are also general purpose but in which the nature of the user's access is the distinguishing characteristic.]

- 2.2.1 Batch systems

    [Jobs are entered locally in batches and processed to completion without user involvement.]

- 2.2.2 Remote job entry systems

    [Jobs may be entered for processing remotely from the central processor, sometimes in batches.]

- 2.2.3 Interactive systems

    [Allow for direct user interaction during the processing of a task.]

    - 2.2.3.1 Time-sharing systems

        [General-purpose interactive systems, capable of handling a variety of problems, languages, etc.]

    - 2.2.3.2 Dedicated applications systems

        [All interaction is concerned with a single application, usually transaction-based (see also 2.3.4.6).]

- 2.2.4 Communications systems

    [Primarily used for transmitting information, with relatively little computation capability at the nodes.]

- 2.2.5 Teleprocessing systems

    [Emphasize user access for interactive computation capability at remote sites.]

2.3 Special Purpose Systems

[Distinguished by the intended class of applications.]

- 2.3.1 Graphics-image processing systems (see also 7.9.2)

    [Emphasis on processing and display of pictorial and graphic information.]

2.3.2 Adaptive systems

[Capable of dynamic modification of the system itself.]

2.3.2.1 Self-organizing systems

[Select and organize their own components or actions to be carried out, based on information about their environments.]

2.3.2.2 Fault-tolerant systems

[Capable of recognizing and recovering from their own hardware failures.]

2.3.3 I/O and memory systems

[Emphasis on input, output and/or storage organization and capability.]

2.3.3.1 I/O control systems

[Intended explicitly for switching and controlling I/O devices.]

2.3.3.1.1 Peripheral subsystems

[Usually organized into clusters of input/output units to facilitate specific functions.]

2.3.3.1.2 Concentrators

[Emphasis is on nodes which collect or distribute streams of input or output.]

2.3.3.2 Hierarchical storage systems

[Organization of storage emphasizes several levels, with differing cost and access characteristics.]

2.3.4 Application-based systems

[Distinguished by type of application.]

2.3.4.1 Air traffic control

[Multiple interactive control of aircraft, usually in the vicinity of an airport, with severe real-time requirements.]

2.3.4.2 Process control

[Usually continuous real-time monitoring and control of an industrial process, such as an oil refinery.]

2.3.4.3 Word processing (see also 8.4)

[Many text processing and retrieval functions.]

2.3.4.4 Message switching

[Primarily transmission and storage of messages over

2. COMPUTER SYSTEMS

communication lines.]

2.3.4.5 Telephone network control

[Control and organization of dynamic paths through a complicated communication network.]

2.3.4.6 Transaction-based

[Many small transactions which may be initiated by many users asynchronously.]

2.3.4.6.1 Reservation

[Interaction with a very rapidly changing data base, and high probability of competitive requests for the same data.]

2.3.4.6.2 Point-of-sale

[Updating a large data base, with many relatively unsophisticated terminals and operators, usually at retail outlets.]

2.3.4.6.3 Electronic funds transfer (see also 9.4.8.3)

[High-volume financial transactions with emphasis on security, reliability and timeliness.]

2.3.4.6.4 Other

[Includes data entry and verification, data correction, etc.]

2.3.4.7 Signal processing

[Emphasis on real-time signal analysis, such as for satellite tracking.]

2.3.4.8 Other

[Includes data base management systems (see 7.4), mathematical modeling systems (see 8.2.3), etc.]

2.4 Performance of Systems

[Measurement of system behavior with respect to normal workload; used to compare systems.]

2.4.1 Attributes

[Characteristics useful for measurement and comparison.]

2.4.1.1 Throughput

[Useful work (other than overhead) performed per unit of time.]

2.4.1.2 Response time

[Time for the system to carry out a task generally regarded as requiring a negligible amount of time and

2. COMPUTER SYSTEMS

communicate back to the user.]

2.4.2 Assessment of attributes

2.4.2.1 Selection studies

[Studies related to the procurement and installation of computer systems and/or related subsystems; usually performed to determine the "best" system among varying alternatives according to a specific performance criterion.]

2.4.2.1.1 Modeling techniques

[An abstract representation of the system is used to measure the behavior of the system or of specific subsystems.]

2.4.2.1.1.1 Simulation models (see also 7.7)

[These models represent the system in the time domain and may be trace-driven (i.e. the input is generated from an actual running system), program-driven (i.e. the input is a description of the system workload) or distribution-driven (i.e. the workload is described by a statistical distribution).]

2.4.2.1.1.2 Analytic models

[These may be probabilistic or deterministic and are solved not by simulation but by some mathematical technique.]

2.4.2.1.2 Measurement techniques

[The actual computer system performance is evaluated by using various tools for this purpose.]

2.4.2.1.2.1 Hardware tools

[Hardware added to a system in order to detect events of interest; normally not accessible to the user.]

2.4.2.1.2.2 Software tools

[Additions to the software system which may record the occurrence of a particular event or which may be used to sample the state of the system at specified time intervals.]

2.4.2.1.2.3 Firmware tools

[Similar to software tools but implemented in microcode so that resolution of what can be measured is finer.]

2.4.2.2 Improvement studies

[These are made in order to determine the effect of possible changes on the performance of a system such as upgrading hardware or changing the scheduler; modeling and measurement techniques used similar to those in selection studies (see 2.4.2.1).]

2.4.2.3 Design studies

[These attempt to deal with performance issues in systems and subsystems which are in the process of being developed; modeling techniques similar to those in selection studies (see 2.4.2.1) are used but, since system may not exist, measurement may not be possible.]

PART IV. ANNOTATED TAXONOMY TREE
3. DATA

3. Data

[Those parts of computer science and engineering which deal with the structuring, storage representation, manipulation and transmission of information.]

3.1 Data Structures

[Refers to how data is organized.]

3.1.1 Primitive

[The smallest logical units of data used by themselves or to construct composite structures; see 3.1.2.]

3.1.1.1 Integer

3.1.1.2 Real

[A number which generally has an integral part and a fractional part; normally expressed in floating-point form (see 5.2.1.2).]

3.1.1.3 Logical

[Refers to an item which can be "true" or "false."]

3.1.1.4 Character

[An item which may be any character from the character set of the computer.]

3.1.1.5 String

[A sequence of characters considered as a single entity in which each character is selected from the character set of the computer.]

3.1.1.6 Pointer

[Refers to an item which specifies the location of another data structure.]

3.1.1.7 Other

[Such as formal algebraic structures and patterns.]

3.1.2 Composite

[Refers to data structures which are constructed from primitive items.]

3.1.2.1 Sets

[An unordered collection of items all of some primitive type.]

3.1.2.2 Arrays

[An ordered sequence of items with some dimensionality; corresponds to a vector in one dimension and a matrix in two dimensions; maximum size in each dimension usually fixed; simplest case is complex number which is sequence of two reals; all elements of array usually of same type but there are exceptions.]

3.1.2.2.1 Indexed access

[Elements in the array are accessed by specifying the indices (or subscripts) which indicate the location of the element in the array (one index for each dimension).]

3.1.2.2.2 Content access

[Elements in the array are accessed by specifying the content (or value) stored in an array location; an example is a Snobol table.]

3.1.2.3 Linear lists

[An ordered sequence of elements whose number may vary with time as items are inserted in or deleted from it.]

3.1.2.3.1 Stack

[A linear list where insertions, deletions on accessing are at one end only and are such that the last inserted item must be the first out (last-in first-out or LIFO).]

3.1.2.3.2 Queue

[A linear list on which insertions and deletions are usually performed only at the ends of the structure (but see 3.1.2.3.2.3).]

3.1.2.3.2.1 FIFO

[A first-in first-out queue where insertions are performed at one end (the rear) and deletions are performed at the other end (the front).]

3.1.2.3.2.2 Deque

[A queue where insertions and deletions can be performed from either end.]

3.1.2.3.2.3 Priority

[A queue in which insertions and deletions can also be made at positions other than the ends typically based on some ranking of the elements.]

3.1.2.3.3 Other

[Such as lists which allow arbitrary

insertion and deletion.]

3.1.2.4 Nonlinear lists

[Lists in which there is no simple ordering among the elements.]

3.1.2.4.1 Associative

[A list in which each element is accessed by naming its contents rather than its position.]

3.1.2.4.2 List structures

[Which may be thought of as lists in which each element may itself be a list ("sublist") and the sublists may themselves have elements which are lists etc.]

3.1.2.5 Trees

[A structure in which each element (node) except the root element is connected by a single branch to its parent and by one or more branches to its children, if any.]

3.1.2.5.1 Binary

[A tree whose nodes have at most two children.]

3.1.2.5.1.1 Heap

[Each leaf (node with no children) has a height (number of branches on path from root to node) of d or d-1 for some d; this term also implies a certain labeling of the nodes to facilitate sorting.]

3.1.2.5.1.2 Height-balanced (AVL)

[For each node the maximum number of branches from the root to a node of the right subtree differs from the height of the left subtree by at most 1.]

3.1.2.5.1.3 Weight-balanced

[For each node, the number of nodes in the right subtree is approximately equal to the number in the left subtree.]

3.1.2.5.2 n-ary

[A tree whose nodes can have at most n children.]

3. DATA

                3.1.2.5.2.1 Trie

                [A complete n-ary tree (each node has 0 or n children) in which typically each of the n children of a node is associated with a unique character from an alphabet.]

                3.1.2.5.2.2 Balanced n-ary tree

                [n-ary trees which are height-balanced.]

    3.1.2.6 Graphs (see 5.2.2)

        3.1.2.6.1 Directed (digraph) (see 5.2.2.1.1)

            3.1.2.6.1.1 Acyclic

            [A graph which contains no closed paths (see also 5.2.2.2).]

            3.1.2.6.1.2 Other

            [Such as arbitrary list structures (see 3.1.2.4.2).]

        3.1.2.6.2 Undirected (see 5.2.2.1.2)

    3.1.2.7 Structures

    [A hierarchy of items in which items at any level can be grouped into a single item at the next level; items may be of different primitive types; actually a tree (see 3.1.2.5) but listed separately to contrast with an array and to emphasize the concept of grouping of items.]

3.1.3 Logical data base

[A structure which contains a set of files in which there are certain associations or relationships between the records of the files.]

    3.1.3.1 Network (see also 7.4.2.1.2)

    [A database structure in which associations are represented by a graph.]

    3.1.3.2 Relational (see also 7.4.2.1.3)

    [A database structure in which associations are represented by a number of interrelated relations.]

    3.1.3.3 Hierarchical (see also 7.4.2.1.1)

    [A database structure in which associations are represented by a set of tree structures.]

3.2 Data Storage Representation

[Refers to the representation of a particular data structure in the memory of a computer; in general, there can be more than one for a particular data structure.]

3.2.1 Primitive items

[The set of primitive structures which are available on most computers; see 3.1.1.]

3.2.1.1 Numeric

[Items which are manipulated in memory according to arithmetic rules.]

3.2.1.1.1 Fixed

[Numbers which are represented and usually interpreted as integers but which may be interpreted as decimal (binary) numbers with the decimal (binary) point in a fixed position in the number.]

3.2.1.1.1.1 Sign-magnitude

[Where a number is represented by its magnitude and a + or - sign.]

3.2.1.1.1.2 2's complement

[Where negative numbers are stored in twos (radix) complement form.]

3.2.1.1.1.3 1's complement

[Where negative numbers are stored in ones (diminished radix) complement form.]

3.2.1.1.2 Floating

[Where numbers are represented in memory by a fractional or integer mantissa and an exponent part.]

3.2.1.1.2.1 Excess notation

[Where the exponent is unsigned (positive) but interpreted to be the true exponent plus (excess) some value so that negative exponents can be represented.]

3.2.1.1.2.2 Multiple-precision

[Where more than the standard single precision word of memory is used to represent a number.]

3.2.1.1.2.3 Other

[Such as signed rather than excess exponents.]

3.2.1.2 Character

[The representation of the characters in a computer's alphabet; see also 3.1.1.4.]

3.2.1.2.1 EBCDIC

[The Extended Binary Coded Decimal Interchange Code designed primarily for IBM machines in which each character is represented by an 8-bit code.]

3.2.1.2.2 ASCII

[The American Standard Code for Information Interchange which was developed as a standard coding scheme for the computer industry; a 7-bit code but also has an 8-bit form.]

3.2.1.3 Pointer

[Where the representation is an address of a memory location; see also 3.1.1.6.]

3.2.1.3.1 Absolute

[A pointer value which is the absolute location at which the data structure resides.]

3.2.1.3.2 Relative

[A pointer value which is an offset relative to some base location where data structures of the type pointed to are located.]

3.2.1.4 Logical

[Where "true" or "false" are typically represented by a "0" or "1" bit; see also 3.1.1.3.]

3.2.2 Record

[A collection of information items about a particular entity.]

3.2.2.1 Fixed length

[All records have the same length.]

3.2.2.2 Variable length

[The length of a record varies from one record to another.]

3.2.3 File

[A collection of records involving a set of entities with

certain aspects in common and organized for some particular purpose.]

3.2.3.1 Internal

[Where the entire file resides in main storage.]

3.2.3.1.1 Sequential

[Where records are stored consecutively in storage.]

3.2.3.1.2 Linked

[Where records need not be stored in consecutive locations but are linked to one another by means of pointers.]

3.2.3.1.2.1 Singly

[The position of the next record is stored in the record which logically precedes it.]

3.2.3.1.2.2 Doubly

[The positions of both the logical predecessor and successor of a record are stored in that record.]

3.2.3.1.2.3 Multilinked

[The positions of many other records may be stored within a record.]

3.2.3.1.3 Hashed

[The location of a record is computed by using an item in the record called a key.]

3.2.3.1.3.1 Division

[Where the computation consists of dividing the key by a quantity and using the remainder as the location of a record.]

3.2.3.1.3.2 Midsquare

[Where the key is multiplied by itself and the middle bits or digits of the result give the desired address.]

3.2.3.1.3.3 Folding

[Where the key is partitioned into a number of parts which are added together to obtain the address.]

3.2.3.1.3.4 Radix transformation

[Where the key represented in one radix is considered to be a number expressed in some other radix for purposes of computing an address.]

3.2.3.1.3.5 Digit analysis

[Where the position of a record is obtained by selecting and shifting digits or bits in its associated key.]

3.2.3.1.3.6 Piecewise linear

[A distribution-dependent method which is based on dividing the key space into a number of equal subintervals.]

3.2.3.1.3.7 Other

[Such as piecewise linear with internal splitting and multiple frequency methods.]

3.2.3.1.4 Other

[Such as files using AVL trees (see 3.1.2.5.1.2) or inverted lists (see 7.3.2.1.3).]

3.2.3.2 External

[Refers to a file which is stored on some secondary storage device.]

3.2.3.2.1 Unordered

[In which the records are not ordered according to any key value.]

3.2.3.2.2 Primary key

[Which allows a record to be located using a single key.]

3.2.3.2.2.1 Sequential

[Where the key value is used to search sequentially through the file to find the desired records.]

3.2.3.2.2.2 Indexed sequential

[Where auxiliary indices are used which give the approximate position of the desired record after which sequential search is used.]

3.2.3.2.2.3 Direct

[Where the key allows the precise position of the record on secondary storage to be computed.]

3.2.3.2.3 Multi-key

[Where more than one key is used to locate a record, one of which is called the primary key while the others are called secondary keys.]

3.2.3.2.3.1 Multi-list

[An organization in which records that have equivalent values for a given secondary key are linked together to form a list.]

3.2.3.2.3.2 Inverted list

[An organization in which for each value of a secondary key a list is maintained of all primary key values of records which contain the secondary key value.]

3.2.4 Physical data base (see also 7.4)

[Refers to the storage of a set of files which exhibit certain associations or relationships between the records of the files.]

3.3 Data Management

[The task of managing internal and external computer storage in order to respond to user requests and to achieve necessary efficiency of use.]

3.3.1 Management disciplines

[Involves responding to requests for storage space as well as the ability to reacquire space which is no longer needed by programs.]

3.3.1.1 Stack-oriented

[Requests for space are handled by taking space from and returning space to the top of an availability stack.]

3.3.1.2 First fit

[Where storage availability is contained in a linear list; storage requests are satisfied by the first block on the list which meets the requirements.]

3.3.1.3 Best fit

3. DATA

[Where the storage is organized as in the first-fit method but the block chosen is that which is closest to the size of the requested block.]

3.3.1.4 Next fit

[Where the available storage is organized as a doubly-linked list of blocks.]

3.3.1.5 Buddy system

[A management technique which arranges available space in pairs of blocks that are called buddies.]

3.3.1.5.1 Binary

[Where each of a pair of buddies is the same size and all block sizes are some power of 2.]

3.3.1.5.2 Fibonacci

[Where the sizes of the two buddies are given by two successive Fibonacci numbers.]

3.3.1.6 Garbage collection

[Methods of returning unused memory to the available storage list.]

3.3.2 Coding

[Techniques for assigning values to items which represent, in a storage-efficient manner, the set of properties that characterize the item.]

3.3.2.1 Fixed length

[Each coded item is represented with the same number of bits or digits.]

3.3.2.1.1 Logical

[An item is encoded with zeros and ones indicating the presence or absence of an attribute.]

3.3.2.1.2 Binary encoded

[An item is encoded as a string of zeros and ones of length $\log_2 n$, where n is the number of unique values that characterize an item.]

3.3.2.2 Variable length

[Coded items are not all represented with the same number of bits or digits.]

3.3.2.2.1 Huffman (see also 5.2.2.4.5)

[A code which minimizes the total storage needed by coding more frequent items with shorter codes.]

3.3.2.2.2 Shannon-Fano

[A variable length coding scheme which uses a table of the frequency of occurrence of items.]

3.3.2.2.3 Tagged

[A coding technique in which the size of a variable length code is explicitly expressed as part of the code.]

3.3.2.2.3.1 Boundary

[Variable length items are bounded by special tag symbols.]

3.3.2.2.3.2 Description

[The length of an item is denoted by a specific descriptor.]

3.3.3 Compaction

[A technique for reducing the size of the physical representation of a set of coded items.]

3.3.3.1 Run-length coding

[A technique which substitutes a count of the run-length and a single instance of an item for a run of identical items.]

3.3.3.2 Pattern substitution

[Frequently occurring patterns or items are replaced with special codes.]

3.3.3.3 Differencing

[Only the difference between contiguous fields in a record or contiguous records is stored.]

3.3.4 Encryption

[Stored data is modified using a transformation algorithm in order to render it incomprehensible to unauthorized examiners.]

3.3.4.1 Transposition ciphers

[The characters comprising the stored data are rearranged but not changed in value.]

3.3.4.2 Standard substitution ciphers

[Such as monoalphabetic or polyalphabetic substitution in which the characters comprising the data are taken in groups of size one or two or more, depending on the system employed, and replaced with values calculated using an algorithm which depends on both the plaintext and a secret key.]

3.3.4.3 Algebraic substitution ciphers

[The numerical equivalents of groups of plaintext characters are transformed to different numbers, thence to cipher text characters, by a numerically oriented algorithm; one example is public key cryptosystems based on mathematical transformations which are easy to apply in one direction (encipherment) but prohibitively time-consuming (for unauthorized eavesdroppers) to apply in the inverse direction.]

3.3.4.4 Hybrid systems

[Plaintext is both transposed and replaced with substitute values; a national Data Encryption Standard specifies one such system for potential widespread use in data storage and communications.]

3.4 Data Communications

[All communications and communication systems for transmitting data between computers or from peripheral or data entry devices to computers.]

3.4.1 Data transmission

[From one device to another without the need for intermediate switching.]

3.4.1.1 Transmission media

[Such as cable (wire-pair, coaxial or optical fibre) or free electromagnetic radiation (video-frequency, microwave or optical).]

3.4.1.2 Modulation and demodulation

[Data to be transmitted is first converted (modulated) to a form more suitable for transmission and then after transmission is converted back (demodulated) to its original form.]

3.4.1.2.1 Modulation techniques

[Such as amplitude, frequency, phase, delta, pulse amplitude, pulse code modulation.]

3.4.1.2.2 Modems

[Acronym for modulator-demodulator which refers to devices for performing the modulation and demodulation such as telephone data sets or acoustic couplers.]

3.4.1.2.2.1 Handshaking procedures

[Connections between two devices which wish to

3. DATA

communicate are established, controlled and, after transmission, broken.]

3.4.1.2.2.2 Modem interface control circuits

[A device which controls the modem to which it is connected; these circuits are the subject of the CCITT standard V24 and the EIA standard RS-232-C.]

3.4.1.3 Multiplexing

[The process by which several lower-speed data streams may be transmitted simultaneously over a single higher-speed line.]

3.4.1.3.1 Space division multiplexing

[Several physically distinct lines are adjacent and may be considered to be a single line.]

3.4.1.3.2 Frequency division multiplexing

[The available frequency range of a communications line is divided into a number of narrower bands, each of which is used for a single, separate data stream.]

3.4.1.3.3 Time division multiplexing

[In which a period of time is divided into several slots each of which is used to transmit a portion of several different data streams.]

3.4.1.3.3.1 Synchronous time division multiplexing

[In which time slots are allotted in a fixed round-robin fashion whether or not a particular channel (data stream) is active.]

3.4.1.3.3.2 Asynchronous time division multiplexing

[Where time slots are allotted only to active channels.]

3.4.1.3.3.3 Contention schemes

[Where a device desiring to transmit must seize a particular time slot not in use.]

3.4.1.3.3.4 Polling schemes

3. DATA

[Where devices which may wish to transmit are polled by the multiplexer to see which wishes to transmit.]

3.4.1.4 Error control

3.4.1.4.1 Sources of errors

3.4.1.4.1.1 Noise

[Any spurious or undesired disturbance that tends to obscure or mask the signal to be transmitted.]

3.4.1.4.1.2 Distortion, attenuation and phase

[Respectively, changes in the form, strength and angular position of a signal.]

3.4.1.4.1.3 Crosstalk

[Errors introduced on one transmission line by signals on a neighboring one.]

3.4.1.4.1.4 Other

[Such as breaks in transmission and echoing.]

3.4.1.4.2 Equalization

[Processes, sometimes called conditioning, by which the static characteristics of a transmission line are adjusted to allow optimal transmission.]

3.4.1.4.3 Coding for error control

[Methods for the detection and correction of transmission errors usually by adding parity or other bits to the message bits.]

3.4.1.4.3.1 Error detecting codes

[Which detect but do not allow the correction of some transmission errors; data in error must be retransmitted.]

3.4.1.4.3.2 Error correcting codes

[Which allow certain detected errors to be corrected without retransmission.]

3.4.1.4.4 Protocols for error control

[The actions taken in order to signal the

occurrence of and to recover from a detected error.]

3.4.1.4.4.1 Negative acknowledgement

[Whereby the receiver of a message signifies that it has been incorrectly received and should be retransmitted by the sender.]

3.4.1.4.4.2 No error acknowledgement

[Whereby the receiver of a message discards any messages received incorrectly; the sender retransmits messages not positively acknowledged after the requisite time-out period.]

3.4.2 Data switching

[The switching of data from one communication path to another; usually needed only when sending and receiving devices are far apart.]

3.4.2.1 Circuit switching

[The connecting of m input lines to n (<m) output trunk lines.]

3.4.2.1.1 Analog switching systems

[A physical path is established between sender and receiver for the duration of the call.]

3.4.2.1.2 Time division switching

[The input lines are sampled sequentially and any change in state transmitted to the output line.]

3.4.2.1.3 Control signaling

[Information is transmitted to enable the appropriate paths to be set up and cleared down.]

3.4.2.2 Store and forward switching

[Where a device, typically a computer, accepts messages from a transmission line, stores them if necessary and then forwards them over a transmission line to the next destination.]

3.4.2.2.1 Message switching

[The entire message is transmitted, stored and forwarded as a single entity.]

3.4.2.2.2 Packet switching

[A message is broken up into (usually equal size) blocks (packets), each of which is transmitted, stored and forwarded separately; different packets may take different paths and may arrive in an order different from the transmission order.]

3.4.3 Economic, legal and regulatory aspects (see also 9.4, 9.7)

3.4.3.1 Privacy and security (see also 9.4.2, 9.4.3)

[Achieved by controlling access to the data communications system by encryption of data and by making the data communications software secure.]

3.4.3.2 Computer-communications interface

[Refers to such considerations as tariff structures, the role of the Federal Communications Commission, specialized versus general common carriers and the role of the computer manufacturer in communications.]

3.4.3.3 International standards

[Communications standards are required to enable interworking between common carriers of different countries and on satellite networks; standards are formulated by bodies such as the International Standards Organization (ISO) and the Comité Consultatiif International Télégraphique et Téléphonique (CCITT).]

PART IV. ANNOTATED TAXONOMY TREE
4. SOFTWARE

4. Software

[All programs, whether designed for a particular application or for usage independent of any application; includes tools, methods and techniques for designing and documenting such programs, and includes the organization of the data used by such programs; contrasts with physical hardware.]

4.1 Tools and Techniques

4.1.1 Programming languages

[A set of characters and rules for combining them which permit communication of programs (including data) from a human to a computer and which do not require specific knowledge of the numerical form of machine instructions or storage locations.]

4.1.1.1 Development

4.1.1.1.1 Design

[Make decisions on purpose, potential users, style and method of definition.]

4.1.1.1.2 Definition techniques

[Notations and other methods used to specify the syntax and/or semantics of programming languages; examples include Backus-Naur Form (=BNF) and Vienna Definition Language (=VDL).]

4.1.1.1.3 Standardization

[The process of formally agreeing upon a single definition of a single language or class of languages.]

4.1.1.1.4 Automatic programming

[The process of using a computer to perform some stages of the work involved in preparing a program; narrow meaning and actual usage of the term changes as the state of the art changes.]

4.1.1.2 Implementation (=Translation)

[How programs written in a programming language are converted to machine code and executed to produce answers.]

4.1.1.2.1 Types of translators (=Processors)

4.1.1.2.1.1 Compiler

[A program which translates source programs (i.e. programs written in a programming language) into a form directly executable by the computer.]

4.1.1.2.1.2 Interpreter

4. SOFTWARE

[A program which examines a source program, interprets its meaning and produces answers directly; it may or may not do a partial translation into some other form.]

4.1.1.2.1.3 Mixed

[A program which contains some features of both compilers and interpreters.]

4.1.1.2.1.4 Cross-compiler

[A compiler which is executed on one computer but produces object code which is to be executed on another.]

4.1.1.2.2 Translator writing techniques

4.1.1.2.2.1 Syntax-directed

[A method whereby the syntactic definition of a programming language is used as input to a compiler generator which produces a compiler specifically designed to translate programs in the language.]

4.1.1.2.2.2 Intermediate language

[A language to which the source program is translated and from which the object program is generated.]

4.1.1.2.2.3 Preprocessor

[A program which translates source programs in one language into another more common language whose translator is usually more readily available.]

4.1.1.2.2.4 Optimization

[Techniques used in a compiler to produce more efficient object code (i.e. directly executable code).]

4.1.1.3 Types (=Classifications)

4.1.1.3.1 By application area

[Languages are usually designed to be primarily useful for specific applications such as numerical computation, business data processing, civil engineering, graphics, simulation (see 7.7.2.3), list processing.]

4.1.1.3.2 Assembly

[A language whose instructions normally correspond to machine instructions on a one-to-one basis, but which allows symbolic names to be used rather than absolute storage locations.]

4.1.1.3.3 Macro assembly

[An assembly language which allows the user to refer to a group of instructions by a single name which can then be used with different values for the variables.]

4.1.1.3.4 High-level

[A term commonly used for languages like Cobol, Fortran, Snobol, Cogo, Simscript to contrast with assembly and machine languages; characteristics of such languages are (1) machine code knowledge unnecessary; (2) good conversion potential; (3) one-many instruction expansion; (4) notation oriented to the problem.]

4.1.1.3.5 Extensible

[Contains a facility as part of the language which allows the user to define new syntax and/or semantics in terms of the existing base language; this facility also causes automatic translation of the new language features to take place.]

4.1.1.3.6 Table/questionnaire

[The user supplies the program in the form of a table (e.g. a decision table), or interacts with the computer to answer questions whose answers cause the program to be executed.]

4.1.1.3.7 Nonprocedural

[A relative term which changes as the state of the art changes; generally applies to a language which allows the user to supply less detail in a program than is common at that point in time; other terms sometimes used in a similar context are higher level, specification, less procedural, goal oriented.]

4.1.1.3.8 Data definition

[A language which is used to define files and/or individual fields in a record.]

4. SOFTWARE

4.1.1.3.9 Other

[Such as publication, hardware, problem defining, special purpose.]

4.1.1.4 Commands (=Statements)

[One of the two major elements in a programming language; see 4.1.1.5 for the other.]

4.1.1.4.1 Control structures (see also 6.3.3)

[The elements in a language which determine the sequence in which actions will be taken.]

4.1.1.4.1.1 Sequential

[Statements, such as assignment statements, which are executed one right after the other.]

4.1.1.4.1.2 Loop

[Provide facilities which allow the user to cause the repetitive execution of a group of statements until some condition holds.]

4.1.1.4.1.3 Conditional

[Provide facilities which allow alternative sequences of execution depending upon the results of certain conditions (e.g. equality of two numbers, availability of data on an input device).]

4.1.1.4.1.4 Unconditional transfer of control

[Specification by the programmer of which statement is to be executed next; usual command name is GOTO.]

4.1.1.4.1.5 Procedure/subroutine invocation

[The method by which a specific portion of code (called a procedure or subroutine) is to be used.]

4.1.1.4.1.6 Parallel control structures

[Statements which specify portions of the program which can be executed in parallel or independently of each other (e.g. fork/join or synchronization statements).]

- **4.1.1.4.2 Storage management**

    [Statements which affect the way in which data is stored in the executable program (e.g. overlays, stacks).]

- **4.1.1.4.3 Input/output**

    [Statements which provide interaction with the external devices (e.g. disks, tapes).]

- **4.1.1.4.4 Error detecting/correcting**

    [Statements which are used to detect and/or correct errors (e.g. testing for division by zero).]

- **4.1.1.4.5 Non-executable**

    [Directives to the translator which do not normally get translated directly into executable code (e.g. comments, file declarations).]

4.1.1.5 Data types and structures

[One of the two major components of a programming language; see 4.1.1.4 for the other.]

- **4.1.1.5.1 Data types (see also 4.1.1.5.3, 3.1.1)**

    [The primitive types of data which can be used in a programming language.]

    - 4.1.1.5.1.1 Arithmetic (see 3.1.1.1 and 3.1.1.2)
    - 4.1.1.5.1.2 Logical (see 3.1.1.3)
    - 4.1.1.5.1.3 Character (see 3.1.1.4)
    - 4.1.1.5.1.4 String (see 3.1.1.5)
    - 4.1.1.5.1.5 Pointer (see 3.1.1.6)
    - 4.1.1.5.1.6 Other

        [Such as procedure variables, label variables and programmer-defined data types.]

- **4.1.1.5.2 Data structures (see also 3.1.2)**

    [Define the ways in which individual data types can be combined in a programming language; normally operations on these data structures are available in the language.]

    - 4.1.1.5.2.1 Arrays (see 3.1.2.2)
    - 4.1.1.5.2.2 Lists (see 3.1.2.3, 3.1.2.4)

4. SOFTWARE

4.1.1.5.2.3 Records (see 3.2.2)

4.1.1.5.2.4 Files (see 3.2.3)

4.1.1.5.2.5 Other

[Such as sets and discriminated unions.]

4.1.1.5.3 Abstract data type

[A data type together with the operators defined on it; the operators can be defined by a user but are independent of any storage representation (e.g. a stack).]

4.1.2 Programming and coding techniques

4.1.2.1 Subroutine (=Procedure)

[A subset of a program which has a use in more than one part of the program, and which can be "called" from different portions of the program; subroutines are normally executed with different values of their parameters and always return control to the part of the program which called them.]

4.1.2.2 Coroutine

[Similar to a subroutine except that a coroutine can call another coroutine before reaching its normal termination and later have control properly returned to the place where it left off.]

4.1.2.3 Recursion

[A technique which allows a program unit (e.g. a procedure) to reference itself.]

4.1.2.4 Reentrant code

[A characteristic of a program which allows multiple simultaneous executions of it.]

4.1.2.5 Error-related

4.1.2.5.1 Debugging techniques

[The collection of programs, methods, and documentation which assist the programmer in checking the correctness of a program and fixing the errors found.]

4.1.2.5.2 Restart procedures

[The collection of programs, methods, computer operator actions, and documentation which make it possible to start a program from the middle if something fails before the program legitimately ends.]

4.1.2.6 Other

[Such as input data checking and checking of data generated during execution.]

4.1.3 Program design and development

4.1.3.1 Requirements

4.1.3.1.1 User

[Requirements which the user needs to have satisfied by the program being developed (e.g. timing).]

4.1.3.1.2 Hardware

[Hardware facilities which may be required by the program being developed (e.g. display equipment, very large memory) or, alternatively, specific hardware which must be used and hence into which the program must fit.]

4.1.3.1.3 Software

[Specifications that the software must satisfy (e.g. fitting into particular hardware, or satisfying timing constraints) or, alternatively, specific software which must be used by and with the program under development and which therefore places constraints upon it.]

4.1.3.2 Specification

4.1.3.2.1 Functional

[The set of capabilities which the program must provide (e.g. calculate a particular trajectory, print pay checks); the functional specifications are often a detailed listing and elaboration of the program requirements; see 4.1.3.1.]

4.1.3.2.2 Performance

[The detailed list of performance capabilities which the program must exhibit; generally these are in terms of timing and memory.]

4.1.3.3 Design techniques

4.1.3.3.1 Simulation/modeling (see also 7.7)

[The use of subsets or approximations to the intended final program for the purpose of investigating performance or trying out alternate programming techniques; simulation and modeling are distinctly different techniques.]

4. SOFTWARE

4.1.3.3.2 Modularity

[The process of preparing a program in small units to make it easier and safer to make changes at any stage in the program development.]

4.1.3.3.3 Top-down/stepwise refinement

[The process of preparing a program by writing down some very very broad functions which the program must perform (e.g. read data, calculate, provide output) and then successively providing more and more details for each of the indicated actions shown in the program.]

4.1.3.3.4 Two-dimensional descriptions

[The use of two-dimensional and/or graphic symbols (e.g. flowcharts) as a means of presenting the program design.]

4.1.3.3.5 Other

[Such as programming support libraries and assertions for program verification.]

4.1.3.4 Documentation (see also 9.5.1.3.4, 9.5.2.2.5 and 9.5.5.4.4)

4.1.3.4.1 Types

4.1.3.4.1.1 Program comments

[Words or symbols embedded in the program but identified as not being part of the program and which are intended to make the program easier to understand.]

4.1.3.4.1.2 Two-dimensional

[Various methods (e.g. flowcharts, logic diagrams) used to describe programs and which use boxes with words and symbols in them.]

4.1.3.4.1.3 Narrative

[A description in natural language of what a program does and how it is organized.]

4.1.3.4.2 Audiences

[The different types of people for whom the documentation is intended to be useful.]

4.1.3.4.2.1 User

4. SOFTWARE

[The person or group which expects either to supply input and/or receive the output of a program.]

4.1.3.4.2.2 Programmer

[The person who is doing maintenance (see 4.1.3.6).]

4.1.3.4.2.3 Operator

[The computer operator who must have specific instructions on how to operate the program and what to do in case something happens either to the hardware or the program itself.]

4.1.3.5 Effectiveness measurements

[The set of tools and techniques for measuring various aspects of the effectiveness of software.]

4.1.3.5.1 Testing

[The process of running the program with sample data to (1) determine whether any errors in the program can be found, and (2) to see how well the requirements and the specifications have been met.]

4.1.3.5.2 Software monitors

[Specific programs whose execution is interspersed with the program being examined in order to analyze the program (e.g. determine which portions are used the most).]

4.1.3.6 Maintenance

[Work done on a program after its first release for normal usage.]

4.1.3.6.1 Repairs/corrections

[Changes made to correct errors which have been found in actual usage.]

4.1.3.6.2 Enhancement

[Improvements made to a program; generally they will be either to provide additional functional capability or to improve performance.]

4.1.3.7 Theoretical

4.1.3.7.1 Program specification

[Specification of the input to and output

from a program using mathematical notation ranging from a relatively informal approach to the strict formalism of mathematical logic.]

4.1.3.7.2 Program verification

[The process of formally (in a mathematical sense) checking that a program does what a formal set of specifications asserts it is supposed to do.]

4.1.4 Portability

[The characteristic of a program which allows it to be run in a different environment than that for which it was originally designed or intended.]

4.1.4.1 Across machines

[The ability to have the program run on more than one machine, and, more importantly, to run on more than one machine <u>family</u>.]

4.1.4.2 Across input/output equipment

[The ability of a program to use physically different input/output equipment than the set for which it was originally designed (e.g. to identify tapes differently, or to use disks instead of tapes).]

4.1.4.3 Across operating systems

[The ability to have the program run under a different operating system than the one for which it was designed, or under which it was debugged.]

4.1.4.4 Across translators

[The ability of a program in a high level language to produce the same results even when it is translated by different compilers or interpreters.]

4.2 Programming Systems

[Those programs which are intended to be used in a variety of contexts and which are independent of any particular application.]

4.2.1 Operating systems

[Programs which control the allocation of hardware resources to individual programs, and which do the scheduling of programs.]

4.2.1.1 Command/job control functions

[The facilities in an operating system which allow the user to indicate what files and data and other software tools or programs are needed to enable the user's program to run.]

4.2.1.2 Time characteristics

[Those aspects of an operating system which pertain to

the clock time involved for running programs.]

4.2.1.2.1 Real time

[Programs which must be executed in a specific amount of time (usually small) because of external factors (e.g. air traffic control; see 2.3.4.1).]

4.2.1.2.2 Interactive (see also 2.2.3)

[The characteristic of an operating system which, for certain types of usage, gives the user quick response and, often, the illusion of being the only user.]

4.2.1.2.3 Batch (see also 2.2.1)

[Programs are grouped together and run at any time which is convenient for the computer center, without regard to specific time constraints.]

4.2.1.3 Elements to be controlled

[The various aspects of the hardware, and programs desiring execution, which must be controlled by the operating system.]

4.2.1.3.1 Input/output

[All elements of the hardware which involve the transmission of information to and from the main memory.]

4.2.1.3.2 Storage allocation

[The control of all aspects of storage usage and access by the operating system.]

4.2.1.3.2.1 Physical

[Control by the operating system of all physical equipment which involves allocation of main and secondary memory.]

4.2.1.3.2.2 Virtual

[A method of handling the physical storage in such a way that the user may write a program as if there were a much larger than actual main memory available.]

4.2.1.3.3 Parallelism

[The ability to run simultaneously as many of the hardware elements as possible; normally this involves I/O running simultaneously with the CPU.]

4.2.1.3.4 Multiprogramming

[A method by which several programs are resident in memory at the same time and executing essentially simultaneously.]

4.2.1.3.5 Multiprocessing (see also 2.1.2)

[Two or more CPUs are linked to the same I/O and/or memory, with all the CPUs under the control of the same operating system.]

4.2.1.3.6 Scheduling of processes

[The control of the order of execution of many processes.]

4.2.1.3.7 Interprocess communication

[The technique by which the operating system ensures that each process receives any necessary inputs from any other process which may be executing.]

4.2.1.4 Protection

4.2.1.4.1 Privacy

[The concept that only authorized users are allowed to see certain programs and/or data.]

4.2.1.4.2 Security

[The set of hardware, software, and management techniques used to ensure privacy.]

4.2.1.4.3 Error handling

[The ability of the operating system, possibly in conjunction with special hardware, to allow software recovery from either hardware or software errors.]

4.2.2 Utilities

[The (usually relatively small) set of routines which perform individual specific tasks for the user and/or for the operating system.]

4.2.2.1 Linkage editor

[A program which links together several programs which may be using the same names for variables and other parameters.]

4.2.2.2 Loader

[A program which provides the absolute storage locations in main memory for a program whose addresses previously were written as relative locations.]

4. SOFTWARE

4.2.2.3 Data transfer

[Programs which transfer data from one storage and/or I/O device to another (e.g. disk to printer).]

4.2.2.4 Debugging tools

[The set of programs to assist the programmer in debugging (see 4.1.2.5.1).]

4.2.2.5 Diagnostic tools

[Similar to debugging tools.]

4.2.2.6 Other

[Such as testing aids, simulators, text editors.]

4.2.3 Human involvement

[The role of the human in using programming systems.]

4.2.3.1 Operators (see also 9.5.5.2)

[The humans responsible for running the computer, normally under instructions from the operating system which the operator can generally override.]

4.2.3.1.1 Scheduling

[The actions taken by the operator to either concur with or override the sequence of program execution given by the operating system.]

4.2.3.1.2 Error recovery

[The actions taken by the operator to either prevent an error, or recover from one made by either the hardware, the software, or the operator.]

4.2.3.1.3 Peripheral manipulation

[The actions taken by the operator to see that the right material is in the right place for use with and by the peripheral equipment.]

4.2.3.2 Data entry (see also 9.5.5.3.1)

[The work done by humans to prepare data in machine readable form.]

4.2.3.2.1 Verification

[Steps taken by humans to ensure that the machine readable data is correct.]

4.2.3.2.2 Correction

[Steps taken by humans to correct machine readable data that has been entered onto

4. SOFTWARE

some device (including e.g. a punched card) once an error has been discovered.]

4.3 Data and File Organization and Management

4.3.1 File organization (see also 3.2.3)

4.3.1.1 Physical

[Involves the way in which data is stored physically on an external storage device such as magnetic tape or disk.]

4.3.1.1.1 Storage media

[The actual type of physical device on which data is stored.]

4.3.1.1.2 Arrangement on storage media

[Involves the way in which the data is physically stored (e.g. with physical gaps between physical blocks).]

4.3.1.1.3 Checking methods (e.g. parity)

[Methods used to check that some hardware error has not caused an error in the data.]

4.3.1.2 Logical

[Refers to the logical or conceptual aspects of the data organization.]

4.3.1.2.1 File description techniques

[Notations and words used to describe files, as contrasted with those describing records and fields.]

4.3.1.2.2 File identification

[Information put at the front and/or the end of a file for identification purposes such as the name of the file, the date created, the number of records.]

4.3.1.2.3 Record layout (see also 3.2.2)

[The organization of specific units of data within a file; within a record are fields which contain detailed information (e.g. street, city, state, and zip code within an address record; each unit is called a field); fields may have subfields.]

4.3.1.2.4 Keys of records (see also 3.2.3.2.2-3)

[The main and subsidiary identifiers of each record such as name or social security number.]

4.3.1.2.5 Indexing and access methods (see also

4. SOFTWARE

3.2.3.2)

[Techniques used to store information on a file and/or in main memory so that the data can be accessed according to some specified rule.]

4.3.2 Data structures and management

[The overall method of organizing data, storing it, controlling access to it, updating it, etc.]

    4.3.2.1 Data description techniques

        [Notations which describe data at all levels, i.e. file, record, field, subfield; differing techniques are needed at different levels.]

    4.3.2.2 Types of structures (see also 4.1.1.5.2 and 3.1.2)

    4.3.2.3 Database management systems (see also 7.4)

        4.3.2.3.1 Database administrator (see 7.4.6.2)
        4.3.2.3.2 Data models (e.g. relational) (see 7.4.2.1)

PART IV. ANNOTATED TAXONOMY TREE
5. MATHEMATICS OF COMPUTING

5. Mathematics of Computing

5.1 Continuous Mathematics

[The underlying mathematical functions are continuous.]

5.1.1 Zeros of nonlinear equations

[Numbers, real or complex, are sought for which a given function or a system of functions vanishes.]

5.1.1.1 Bisection

[A simple method for finding a real isolated zero of a function $f(x)$ by successively halving an interval in which the zero is known to lie.]

5.1.1.2 Fixed point iteration

[A method which starts with a given approximation x_0 to a zero of a function and attempts to obtain improved approximations using an iteration $x_{n+1}=g(x_n)$.]

5.1.1.3 Secant method

[Starts with two approximations x_0 and x_1 to a zero of a function $f(x)$ and selects as the next approximation x_2 the intersection of the x-axis with the secant through the points $(x_0,f(x_0))$ and $(x_1,f(x_1))$.]

5.1.1.4 Newton's method

[An iterative method which, starting with a given approximation x_0, computes successive approximations by $x_{n+1}=x_n- f(x_n)/f'(x_n)$.]

5.1.1.5 Muller's method

[An iterative method analogous to the secant method (see 5.1.1.3) in that three approximations are used at each stage to compute the intersection of the x-axis with a quadratic passing through three points.]

5.1.1.6 Systems of equations

5.1.1.6.1 Newton-based

[Generalizations of Newton's method (see 5.1.1.4) for a single equation to the greater than one equation case.]

5.1.1.6.2 Generalized secant

[A corresponding generalization of the secant method (see 5.1.1.3).]

5.1.1.6.3 Gradient

[Methods based on finding the minimum of a function of several variables.]

5.1.1.7 Special methods for polynomials

[Such as those which obtain all the zeros simultaneously or which promote efficient calculation of complex roots; an example is the quotient-difference algorithm.]

- 5.1.1.8 Global convergence techniques

    [Methods that can be used to find good initial estimates of a zero or zeros of a function.]

5.1.2 Numerical integration and differentiation

- 5.1.2.1 Finite difference formulas for differentiation

    [Methods based on using the differences of values of functions, e.g., $f'(x_0) \approx ((f(x_0+h)-f(x_0-h))/2$.]

- 5.1.2.2 Numerical integration - equal intervals

    [The calculation of approximate values of a definite integral at equally spaced points.]

    - 5.1.2.2.1 Trapezoidal rule

        [The area is approximated by a series of trapezoids inscribed between the function curve and the x-axis.]

    - 5.1.2.2.2 Simpson's rule

        [A generalization of the trapezoidal rule in which sections of the curve to be integrated are approximated by quadratic polynomials instead of straight lines.]

    - 5.1.2.2.3 Newton-Cotes formulas

        [A generalization of Simpson's Rule which approximates the curve to be integrated by polynomials of higher degree.]

- 5.1.2.3 Euler-Maclaurin formulas

    [Generalizations of the trapezoidal rule which incorporate correction terms.]

- 5.1.2.4 Romberg integration

    [A technique for improving the order of accuracy of the trapezoidal rule using a process of successive extrapolation.]

- 5.1.2.5 Gaussian quadrature

    [A method for numerical integration in which the abscissa values are not equally spaced but are chosen to achieve maximum accuracy.]

- 5.1.2.6 Adaptive quadrature

    [A procedure which provides automatically for a refined mesh on those sections of an interval where the function requires finer spacing for the desired accuracy.]

5.1.3 Methods for ordinary differential equations

5.1.3.1 One step methods

[Calculate the approximation to the solution of y'=f(x,y) at a sequence of points using only the value of x and y at a previous point.]

5.1.3.1.1 Taylor series

[Approximate values at a sequence of points are calculated using a Taylor expansion about a given initial point.]

5.1.3.1.2 Runge-Kutta

[Calculates values of y at successive points x_i using the value at the previous point but requiring evaluation of f(x,y) for values of x between the two points.]

5.1.3.1.3 Runge-Kutta Fehlberg

[A Runge-Kutta method which also provides an error estimate at each integration step.]

5.1.3.2 Multistep methods

[Require the value of x and y at several, usually equally spaced, points in order to continue the integration.]

5.1.3.2.1 Predictor-corrector

[A combination of two multistep methods which give two estimates of the value of y at a point x_i and from these an estimate of the error is obtained.]

5.1.3.2.2 Variable-order-variable-step

[Allows both the order of the method being used and the size of the next integration step to be changed repeatedly.]

5.1.3.3 Extrapolation methods

[Use extrapolation of the results of numerical integration at two different step sizes to improve the accuracy of a given method.]

5.1.3.4 Stability

[The study of how errors introduced at one step of a method propagate into succeeding steps.]

5.1.3.5 Convergence

[The study of whether or not the numerical solution converges in theory to the true solution as the step size goes to zero.]

5. MATHEMATICS OF COMPUTING

5.1.3.6 Methods for stiff equations

[A study of special methods for integrating differential equations which describe processes with widely varying time scales.]

5.1.3.7 Boundary value problems

[Differential equations with conditions specified at more than one point.]

5.1.3.7.1 Shooting methods

[Guesses of missing conditions at one (initial) point are made and improved by iteration.]

5.1.3.7.2 Finite difference methods

[Discretize the differential equations at a set of mesh points.]

5.1.3.7.3 Superposition methods

[For linear systems the desired solution is obtained by taking a combination of fundamental solutions.]

5.1.3.7.4 Collocation

[Uses a linear combination of N basis functions in the form

$$y(x) = \sum_{k=1}^{N} \alpha_k w_k(x)$$

to obtain an approximation to the solution $y(x)$ of the boundary value problem which satisfies the differential equation at a set of interior points.]

5.1.3.7.5 Finite element methods (see also 5.1.4.1.2 and 5.1.4.2.5)

[Uses basis elements defined at a set of knot points to approximate a desired solution.]

5.1.3.7.6 Variational methods

[Determine the weights in the approximation $y(x) = \sum \alpha_k w_k(x)$ by minimizing a functional which is related to the boundary value problem; the energy functional, used in the Rayleigh-Ritz method, is commonly used.]

5.1.3.7.7 Invariant imbedding

[Converts a boundary value problem into an equivalent initial value problem.]

5.1.3.7.8 Continuation methods

[Solves a sequence of simpler boundary value problems whose limit is the required boundary value problem.]

5.1.4 Methods for partial differential equations

5.1.4.1 Elliptic equations

[Equations defined on a closed region in two or more independent variables.]

5.1.4.1.1 Finite difference methods (see also 5.1.2.1)

[Replaces differential operators by discrete operators defined on a mesh over the region of interest.]

5.1.4.1.2 Finite element methods (see also 5.1.3.7.5 and 5.1.4.2.5)

[Use basis functions defined on a discrete mesh to approximate the differential operators.]

5.1.4.1.3 Gauss-Seidel and overrelaxation methods

[Iterative methods for solving the linear algebraic systems that result from applying finite difference methods.]

5.1.4.1.4 Alternating direction implicit method

[A special iterative method applicable to certain classes of elliptic problems.]

5.1.4.1.5 Fast Poisson solvers

[Direct methods for solving sparse (i.e., the matrix of coefficients has few nonzero elements) linear systems that arise from finite difference approximations, especially applicable to a class of equations studied by the mathematician Poisson.]

5.1.4.1.6 Galerkin's method

[Converts a partial differential equation to a system of ordinary differential equations by expansion of all but one coordinate into a set of orthogonal functions.]

5.1.4.2 Initial value problems (see also 5.1.4.1.1)

[Equations involving one time-like variable and one or more space variables - usually of parabolic or hyperbolic type.]

5.1.4.2.1 Finite difference methods (see 5.1.4.1.1)

5.1.4.2.2 Method of lines

[The equation is discretized in one variable but not discretized in the other variables.]

5.1.4.2.3 Method of characteristics

[The partial differential equation is converted to ordinary differential equations on curves called characteristics; applicable primarily to hyperbolic equations.]

5.1.4.2.4 Convergence and stability (see 5.1.3.4 and 5.1.3.5)

5.1.4.2.5 Finite element methods (see also 5.1.3.7.5 and 5.1.4.1.2)

[Basis elements are used to approximate the solution in one or more of the independent variables.]

5.1.5 Integral and integro-differential equations

[Equations of the form $\lambda x(s) - \int_a^b K(s,t)x(t)\,dt = y(s)$ where $K(s,t)$ and $y(s)$ are given functions and $x(s)$ must be determined on the interval $[a,b]$.]

5.1.5.1 Method of successive approximations

[Starts with some initial guess $x_0(s)$ and generates improved approximations according to prescribed rules.]

5.1.5.2 Interpolatory methods

[Replaces $K(s,t)$ by a piecewise polynomial interpolant defined on subdivisions of the interval $[a,b]$ after which direct integration is possible.]

5.1.5.3 Orthonormal expansions

[$K(s,t)$ is expanded into a series of predetermined orthogonal functions followed by direct integration leading to a system of linear equations.]

5.1.5.4 Collocation method (see also 5.1.3.7.4)

[Assumes an expansion for $x(s)$ in the form $\sum_{i=1}^{N} \alpha_i \phi_i(s)$ where $\phi_i(s)$ are a set of usually orthogonal functions and the α_i are constants which must be determined.]

5.1.5.5 Galerkin's method (see also 5.1.4.1.6)

[Similar to collocation (see 5.1.5.4) but uses a different method to determine the coefficients α_i.]

5.1.5.6 Methods based on numerical integration

[Here the $\int_a^b K(s,t)dt$ is approximated by any standard numerical integration method, e.g. Gaussian quadrature (see 5.1.2.5) and a solution is obtained on a set of node points.]

5.1.6 Interpolation and approximation

5.1.6.1 Interpolation

[The process of finding simple functions, such as polynomials, which agree with a given function and possibly its derivatives at a given set of points.]

5.1.6.1.1 Lagrangian

[The interpolating function is a polynomial whose values match exactly those of the given function at a set of points.]

5.1.6.1.2 Newton form of the interpolating polynomial

[Another form of the interpolating polynomial based on the use of divided differences (see 5.1.6.1.3).]

5.1.6.1.3 Divided differences

[A special notation which leads to an efficient and useful way to derive the Newton form of the interpolating polynomial, e.g. $f[x_1,x_2] = (f(x_2)-f(x_1))/(x_2-x_1)$.]

5.1.6.1.4 Hermite

[Polynomial interpolation in which the interpolating polynomial agrees with a given function in both value and slope at a set of points.]

5.1.6.1.5 Spline

[Polynomial interpolation which maintains certain continuity requirements on the function and its derivatives at selected points (knots).]

5.1.6.1.6 Trigonometric

[Interpolation using trigonometric functions instead of polynomials.]

5.1.6.2 Approximation

[The process of approximating a given function by a simpler class of functions.]

5.1.6.2.1 Least squares polynomial

[Minimizes the sum of the squares of the difference between a given function and an approximating polynomial.]

5. MATHEMATICS OF COMPUTING

5.1.6.2.2 Fourier least squares

[Minimizes the sum of the squares of the difference between a given function and an approximating sum of trigonometric functions.]

5.1.6.2.3 Fast Fourier transform

[A technique for obtaining a trigonometric approximation which minimizes the computer time required to compute the coefficients.]

5.1.6.2.4 Minimax polynomial

[A method which minimizes the maximum of the absolute value of the difference between a given function and an approximating polynomial.]

5.1.6.2.5 Rational

[Uses rational functions, i.e. ratios of polynomials, to approximate a given function.]

5.1.6.2.6 Spline (see 5.1.6.1.5)

5.1.6.2.7 For elementary functions

[Special approximations used to compute accurately and efficiently elementary functions such as sin x, e^x etc. for use in subroutine libraries.]

5.1.6.2.8 Data analysis and smoothing

[The approximating functions, usually polynomials, are used to adjust the values of empirical data in order to reduce the effect of errors in the data after which the data is analyzed.]

5.1.7 Optimization

[The process of finding the minimum or the maximum of a function of several variables.]

5.1.7.1 Least squares

5.1.7.1.1 Linear

[A set of N empirical data points (x_i, y_i) is fitted by a linear combination

$$\sum_{j=1}^{p} \alpha_j \phi_j(x)$$

where $\phi_j(x)$ are given and α_j are determined by minimizing

$$\sum_{i=1}^{N} (y_i - \sum_{j=1}^{p} \alpha_j \phi_j(x_i))^2.]$$

5.1.7.1.2 Nonlinear

[Same as linear least squares except that the approximating function is nonlinear.]

5.1.7.2 Minimization of nonlinear functions

5.1.7.2.1 Newton's method

[Reformulates the problem as one requiring the finding of zeros of a system of nonlinear equations and then using Newton's method (see 5.1.1.4).]

5.1.7.2.2 Quasi-Newton methods

[Newton-like methods which reduce the computational effort required.]

5.1.7.2.3 Gradient methods

[Methods which attempt to follow the path of steepest descent along a surface to its minimum.]

5.2 Discrete Mathematics

[Where the mathematical objects are discrete and/or where the mathematical processes proceed by discrete steps.]

5.2.1 Computer arithmetic

[Arithmetic and number systems used by or inherent to computers.]

5.2.1.1 Roundoff error

[The error due to taking a real number with an infinite decimal representation and truncating it to a number with a finite representation.]

5.2.1.2 Floating-point arithmetic

[A number system used on many computers which automatically adjusts exponents, analogous to scientific notation $a \times 10^b$.]

5.2.1.3 Significant digit arithmetic

[A system designed to keep track of the digits of accuracy lost due to roundoff or due to other loss of significance errors.]

5.2.1.4 Interval arithmetic

[A system which attempts to produce at every stage of a computation an interval within which the true value of a quantity must lie regardless of roundoff errors in arithmetic.]

5.2.1.5 Radix number systems

[Number systems with a base (or radix) which is 10 in

the decimal system; binary (radix 2), octal (radix 8) and hexadecimal (radix 16) are commonly used in computers.]

5.2.2 Graph theory

[A graph consists of a set of points (nodes, vertices), pairs of which may be joined by edges (branches).]

5.2.2.1 Basic concepts

5.2.2.1.1 Directed graphs

[Graphs in which each edge joining two vertices has an explicit direction; there may be edges in both directions between two vertices.]

5.2.2.1.2 Undirected graphs

[Edges between vertices do not have indicated directions.]

5.2.2.1.3 Operations on graphs

[Such as forming the intersection, union, product, composition, partition of graphs.]

5.2.2.1.4 Domination and independence

[A subset of the vertices of a graph is dominating if all vertices not in the set are adjacent (i.e. are joined by edges) to vertices in the set; a set of vertices is independent if no pair in the set is adjacent.]

5.2.2.1.5 Linear graphs

[Graphs with at most one edge joining two vertices; also called simple graphs.]

5.2.2.1.6 Matching and factors

[A k-factor is a spanning subgraph (<u>see also</u> 5.2.2.4.4) where the degree of (i.e. the number of edges incident on) each of its vertices is k; a 1-factor is called a complete matching.]

5.2.2.2 Paths and circuits

[A path is a sequence of edges in which the terminal vertex of each edge, except possibly the last one, coincides with the initial vertex of the next edge; a circuit (or cycle) is a path which begins and ends at the same vertex.]

5.2.2.2.1 Connectivity

[In a connected graph there exists a path between any two vertices.]

5.2.2.2.2 Eulerian path and circuit

[A path and circuit that traverse every edge in a graph exactly once.]

5.2.2.2.3 Hamiltonian path and circuit

[A path and circuit that traverse every vertex in a graph exactly once.]

5.2.2.2.4 The shortest path problem

[With each edge assigned a weight (or distance), refers to finding a minimum distance path between two given vertices in a graph.]

5.2.2.2.5 The travelling salesman problem

[Finding a minimum distance Hamiltonian circuit in a graph.]

5.2.2.2.6 Transitive closure

[The representation of partially ordered sets (posets) by graphs and the transitivity property of the relation on the poset.]

5.2.2.3 Planar graphs

[Which can be drawn in two dimensions so that no two edges intersect.]

5.2.2.3.1 Euler's formula

[$E=V+R-2$ where E, V and R represent, respectively, the edges, vertices and regions of a planar graph.]

5.2.2.3.2 Kuratowski's theorem

[Characterizes planar graphs as not having subgraphs of two specific kinds.]

5.2.2.3.3 Dual graphs

[A relation between two graphs such that every circuit in one corresponds to a cut set (see 5.2.2.4.7) in the other and vice versa.]

5.2.2.3.4 Genus, thickness, coarseness, crossing number

[Genus - number of "handles" on a sphere to embed the graph; thickness - minimum number of planar graphs into which the graph can be decomposed; coarseness - maximum number of edge - disjoint subgraphs; crossing number - number of edge crossings when graph drawn in the plane.]

5. MATHEMATICS OF COMPUTING

5.2.2.3.5 Colorability and chromatic number

[A k-colorable graph allows each vertex to be colored with one of k colors so that no adjacent vertices have the same color; the chromatic number of a graph is the minimum k for which it is k-colorable.]

5.2.2.3.6 The four-color theorem

[Recently proved classical theorem which states that four colors are sufficient to color all planar graphs so that no two vertices of the same color are joined by an edge.]

5.2.2.4 Trees

[Directed graphs with no circuits, a distinguished vertex (the root) with no incoming edges and with all other vertices having a single incoming edge.]

5.2.2.4.1 Characterization of trees

[Concerned with various equivalent definitions and representations of trees.]

5.2.2.4.2 Free trees and centroids

[A free tree is a connected and undirected graph with no cycles; a centroid is any point which, if taken as the root of a tree, results in minimum maximum distance of the root from all other vertices.]

5.2.2.4.3 Enumeration of trees

[Counting the number of trees of a given type such as all binary trees (in which each vertex has at most two outgoing edges) with n nodes.]

5.2.2.4.4 Minimum spanning trees

[Algorithms for finding a tree whose vertices contain all those of a given graph and with a minimal number of edges or total weighted length of edges.]

5.2.2.4.5 Huffman trees

[Trees related to optimal prefix codes, useful in search and merge procedures (see also 3.3.2.2.1).]

5.2.2.4.6 Search trees

[Trees, such as binary search trees, which are related to the searching of records in a file.]

5.2.2.4.7 Cut-sets

[A minimal set of edges whose removal from a connected graph separates it into two disconnected components.]

5.2.2.4.8 Network flow problems

[Problems concerned with the flow of commodities in transportation networks.]

5.2.3 Combinatorics

[Is concerned with arrangements, selections and operations in a discrete system.]

5.2.3.1 Permutations and combinations

5.2.3.1.1 Permutations

[The number of ordered arrangements of a set of objects.]

5.2.3.1.2 Combinations

[The number of ways r objects may be selected from n objects.]

5.2.3.1.3 Binomial coefficients and Stirling numbers

[A binomial coefficient is the coefficient of x^r in the expansion of $(1+x)^n$; if n is an integer it equals $n!/(r!(n-r)!)$ and expresses the number of combinations of n things r at a time; Stirling numbers are related to the ways to partition sets into subsets and the number of permutations of sets with cycles.]

5.2.3.1.4 Generation of permutations and combinations

[Concerned with algorithms for generating permutations and combinations.]

5.2.3.2 Recurrences

[Relationships of the form $r_n = f(r_{n-1}, r_{n-2}, \ldots, r_{n-k})$ for some function f.]

5.2.3.2.1 Linear recurrences

[The function f has the form $a_{n-1}r_{n-1} + a_{n-2}r_{n-2} + \ldots + a_{n-k}r_{n-k}$ and the a_j are (usually) constants.]

5.2.3.2.2 Nonlinear recurrences

[The function f may contain products or other nonlinear functions of the r_i.]

5.2.3.3 Generating functions

[Functions $F(x)$ which, when expanded in powers of x, have a_i or some function of a_i as the coefficient of

5. MATHEMATICS OF COMPUTING

x^i for a given sequence $a_0, a_1, a_2, \ldots$.]

5.2.3.3.1 Ordinary generating functions

[Here a_i is exactly the coefficient of x^i.]

5.2.3.3.2 Exponential generating functions

[Here $a_i/i!$ is the coefficient of x^i.]

5.2.3.4 Principle of inclusion and exclusion

5.2.3.4.1 General formula

[Corresponds to the principle which governs the number of elements included in or excluded from a set of subsets of n objects.]

5.2.3.4.2 Mobius inversion

[An extension of the principle which applies to partially ordered sets.]

5.2.3.5 Polya's theory of counting

[Methods for enumerating nonequivalent objects in sets.]

5.2.3.6 Block designs

[Methods for selecting subsets (blocks) of a given set which satisfy prescribed conditions (e.g. that every object should appear in the same number of blocks).]

5.2.3.7 Ramsey theory

[Relates the number of cliques and independent sets in a graph to the number of vertices in the graph.]

5.2.4 Linear equations and linear algebra

[Concerned with solutions of the system $A\underline{x}=\underline{b}$ where A is a (usually) square matrix, with solutions of the eigenvalue problem $A\underline{x}=\lambda\underline{x}$ or its generalizations and with the calculation of determinants.]

5.2.4.1 Direct methods for linear systems

[Methods which theoretically produce the exact solution of $A\underline{x} = \underline{b}$ in a finite number of steps.]

5.2.4.1.1 Gaussian elimination

[Reduces a given linear system $A\underline{x}=\underline{b}$ to an equivalent upper triangular system.]

5.2.4.1.2 Gauss-Jordan reduction

[Reduces a given linear system to an equivalent diagonal system.]

5.2.4.1.3 Factorization methods

[Methods for decomposing a matrix A into the product of a lower triangular matrix and an upper triangular matrix.]

5.2.4.1.4 Sparse matrix methods

[Take advantage of a large number of zero coefficients in A to reduce the computer time.]

5.2.4.1.5 Methods for inverses and pseudoinverses

[These find A^{-1} when A is square and nonsingular and find a pseudoinverse (or generalized inverse) when A is singular or not square.]

5.2.4.2 Conditioning and error analysis

[Conditioning refers to adjustments of the coefficients of A which reduce the effects of roundoff error; error analysis attempts to find estimates of or bounds on this error.]

5.2.4.3 Iterative refinement

[A method for improving the accuracy of a computed solution, used particularly when serious errors from roundoff are expected.]

5.2.4.4 Iterative methods

[Methods designed to produce an approximation to a solution by iteration starting from some initial guess.]

5.2.4.4.1 Jacobi

[The values of all elements of $\underline{x}$ are changed simultaneously, the ith equation being used to calculate a new value of x_i.]

5.2.4.4.2 Gauss-Seidel

[The ith equation is used to calculate a new value of x_i using the most recently computed values of all x_j, $j \neq i$.]

5.2.4.4.3 Successive overrelaxation

[The equations of a method like Gauss-Seidel are adjusted using an overrelaxation factor in order to accelerate convergence of the method.]

5.2.4.4.4 Conjugate gradient

[A finite iterative method which, starting from an initial approximation $\underline{x}_0$, chooses n directions and distances in Euclidean n-space to find the solution of n equations.]

5. MATHEMATICS OF COMPUTING

5.2.4.4.5 Other

[Such as steepest descent and ordinary relaxation.]

5.2.4.5 Eigenvalues

[The values of λ and x which satisfy $A\underline{x}=\lambda\underline{x}$ or $A\underline{x}=\lambda B\underline{x}$ where B is a matrix.]

5.2.4.5.1 Power iteration

[Usually used to find a dominant (maximum) eigenvalue of a matrix by computing successive powers $A^n\underline{v}_0$ of a starting vector $\underline{v}_0$.]

5.2.4.5.2 Jacobi method

[Used to find all the eigenvalues of a (usually) symmetric matrix by reducing a given matrix to diagonal form by successive reduction of off-diagonal elements to zero.]

5.2.4.5.3 Householder's and Givens' methods

[Reduce a symmetric matrix to tridiagonal (i.e. only the main diagonal and its adjacent two diagonals are nonzero) form from which its eigenvalues are relatively easily calculated; also reduces a nonsymmetric matrix to Hessenberg form (i.e. only the upper triangle and the diagonal below the main diagonal are nonzero) from which other methods (see 5.2.4.5.4) can be used for the eigenvalues.]

5.2.4.5.4 Q-R method

[Transforms a matrix A which has usually already been reduced to Hessenberg form into upper triangular form from which the eigenvalues can be obtained.]

5.2.4.5.5 Lanczos' algorithm

[Another method for transforming a matrix to tridiagonal form using orthogonal transformations.]

5.2.4.5.6 Singular value decomposition

[Orthogonal transformations are used to decompose A into a diagonal matrix whose elements are the singular values (related to the eigenvalues of AA^T) of A.]

5.2.4.6 Determinants

[A numerical value assigned to a square matrix; used, for example, in Cramer's rule to solve linear systems.]

5.2.4.6.1 Expansion by cofactors

[A method to compute the value of a determinant based on an expansion of the form

$$\det(A) = \sum_{i,j=1}^{n} a_{ij} C_{ij} \text{ where } C_{ij} = (-1)^{i+j} \det(A_{ij})$$

and C_{ij} is the cofactor of a_{ij}.]

5.2.4.6.2 By Gaussian elimination

[Quantities calculated during Gaussian elimination (see 5.2.4.1.1) are used to calculate the determinant.]

5.2.5 Mathematical programming

[Techniques used to find the optimum value of an objective function subject to constraints, usually inequalities.]

5.2.5.1 Linear programming

[Applies to problems where both the objective function to be minimized and the constraints are linear.]

5.2.5.1.1 Simplex method

[Adjacent vertices of the convex polygon on which the solution must lie are successively tested until the solution is found.]

5.2.5.1.2 Other

[Including variations of the simplex method and other optimization techniques.]

5.2.5.2 Nonlinear programming

[Applies to problems where either the objective function or the constraints are nonlinear, most commonly quadratic.]

5.2.5.3 Integer programming

[Mathematical programming problems where the variables can only take on integer values.]

5.2.6 Mathematical statistics and probability

5.2.6.1 Random number generators

[Methods to generate pseudo-random numbers, that is numbers generated by a deterministic process but which exhibit the characteristics of random numbers.]

5.2.6.1.1 Linear congruential generators

[The most common method in which the (n+1)st number r_{n+1} is generated by the formula $r_{n+1} = (ar_n + c)$ modulo m where all quantities are integers.]

5. MATHEMATICS OF COMPUTING

5.2.6.1.2 Normal random numbers

[Uniformly distributed numbers generated by the linear congruential method are transformed to exhibit the characteristics of the normal distribution.]

5.2.6.1.3 Other frequency distributions

[Uniformly distributed numbers are transformed to exhibit the characteristics of other frequency distributions such as the Poisson distribution.]

5.2.6.2 Monte Carlo methods

[Sampling techniques using randomly generated points are used to calculate the values of functions or the solution of equations.]

5.2.6.2.1 For integration in multi-dimensional spaces

[Points are chosen randomly in multi-dimensional space and an approximation to the integral calculated by determining the proportion of points within the volume defined by the integrand.]

5.2.6.2.2 For integral equations (see 5.1.5)

5.2.6.3 Probability distributions

[Such as the exponential, Poisson or hypergeometric which are especially useful in simulating job streams in a computer environment.]

5.3 Numerical Software and Algorithm Analysis

[Numerical software refers to computer software packages to solve numerical problems in which the properties and behavior of the algorithms used are well understood in relation to the class of problems to which the package is to be applied.]

5.3.1 Algorithm selection factors

5.3.1.1 Rates of convergence

[Some algorithms converge more rapidly than others.]

5.3.1.2 Efficiency

[The total computational effort required to produce a solution.]

5.3.1.3 Error analysis and stability

[These factors effect the accuracy of the solution and must be considered in algorithm selection.]

5.3.1.4 Memory constraints

[Algorithms differ in the amount of high speed memory

5. MATHEMATICS OF COMPUTING

required and hence some algorithms may be more suitable for a given computer and memory than others.]

5.3.1.5 Computer architecture constraints

[Algorithms are sometimes written to take advantage of the architecture of a particular machine, e.g. parallel computers require specially designed algorithms for maximum efficiency.]

5.3.2 Numerical software

5.3.2.1 Portability constraints (see also 4.1.4)

[The extent to which software written to run on one computer can run successfully on other computers.]

5.3.2.2 Effect of languages and compilers

[Both the language used and the language translator affect the quality of the software.]

5.3.2.3 Testing and certification

[The testing of a software package to eliminate all reasonable errors that might be encountered in solving a specified class of problems.]

5.3.2.4 Program verification

[Various techniques, theoretical or experimental, for verifying that under proper input the software will produce correct answers.]

5.3.2.5 Reliability and robustness

[A program is reliable and robust to the extent that it usually behaves as it is supposed to under a wide selection of problems and input, and even when it does fail, it provides usable diagnostics.]

5.3.2.6 Problem-oriented languages and systems

[Languages specifically designed to address a certain class of problems, e.g. **Cogo** is a language designed to make it easy for Civil Engineers to write programs to solve their problems.]

PART IV. ANNOTATED TAXONOMY TREE
6. THEORY OF COMPUTATION

6. Theory of Computation

6.1 Switching and Automata Theory

[The theory covering electrical and electronic switching circuits and the study of abstract, mathematically idealized machines containing such circuits which are called automata.]

6.1.1 Switching theory

6.1.1.1 Boolean algebras

[Algebras where the values of constants, variables and functions are 0 or 1.]

6.1.1.2 Synthesis of combinational circuits

[Design of circuits without delays (i.e. without memory).]

6.1.2 Sequential machines

[Devices with a finite-state memory, which change states and produce outputs according to predetermined rules as a function of their present states and inputs.]

6.1.2.1 Input/output conventions

[Describe the timing and manner in which outputs are produced.]

6.1.2.2 State reduction

[An algorithm for finding a machine with equivalent behavior and a minimal number of states.]

6.1.2.3 Identification experiments

[Determining information about the present state of a machine by giving it inputs and analyzing the outputs obtained.]

6.1.2.4 Information lossless machines

[Machines having the property that, by observing only the outputs, the input sequence can be fully determined.]

6.1.2.5 Neural nets

[Networks of idealized neurons, which fire synchronously according to specific criteria; equivalent in capabilities to sequential machines.]

6.1.3 Finite-state acceptors

[Sequential machines with 0-1-valued outputs, with the convention that an input string is accepted only if the last output digit is a 1.]

6.1.3.1 Closure properties

[Properties of the classes of sets of strings

(languages) defined by finite-state acceptors such as whether these classes are closed under operations such as set union, complement, etc.]

6.1.3.2 Nondeterministic machines

[For a given state and input, a set of possible next states is specified rather than a single one; see also 6.1.4.3.]

6.1.3.3 Regular expressions

[String expressions which consist of individual string variables and constants formed using the operations of union, concatenation and iteration (closure).]

6.1.3.4 Decision properties

[Properties which are decidable, given the description of a finite-state acceptor (e.g. whether a given machine accepts an infinite set of strings).]

6.1.4 Generalized finite-state machines

[Devices having a finite-state memory but with variations on the basic mode.]

6.1.4.1 Two-way automata

[Machines having the input on a nonerasable two-way tape so that the input can be scanned more than once.]

6.1.4.2 Multitape automata

[Machines having more than one input tape; the tape to be read from at each step is determined by the present state of the machine.]

6.1.4.3 Probabilistic automata

[Machines in which the state transitions occur in a probabilistic manner rather than being fully determined by the present state and input; see also 6.1.3.2.]

6.1.4.4 Tree automata

[Machines having trees of symbols as inputs instead of strings.]

6.1.5 Infinite-state machines

[Devices having unbounded memory capacity.]

6.1.5.1 Counter machines

[Have a finite-state memory plus a counter; the latter can be increased or decreased by one or tested for zero on a given step.]

6.1.5.2 Pushdown automata

[Have a finite-state memory plus a pushdown (last-in first-out) store in which only the top element of the store can be accessed.]

6.1.5.3 Stack automata

[Pushdown automata with the additional property that the contents of the pushdown store can be read (but not altered).]

6.1.5.4 Turing machines (see 6.4.1)

6.1.5.5 Iterative arrays of finite automata (see 6.4.3)

6.2 Formal Languages

6.2.1 The Chomsky hierarchy

[The first and still standard taxonomy of formal languages based on generating systems, or grammars].

6.2.1.1 Unrestricted (Type 0) grammars

[These grammars allow rules that rewrite portions of already generated strings in any way, using two-sided context; e.g., an acceptable rule might read: when the string x is flanked by the string on the right and the string on the left, then x can be rewritten as the string y (where y may be the null string); type 0 grammars generate the class of languages accepted by Turing machine acceptors; see 6.4.1.]

6.2.1.2 Context-sensitive (Type 1) grammars

[These grammars are basically Type 0 grammars in which a string x can be rewritten only by a string y that is at least as long as x; in particular, no erasing is allowed.]

6.2.1.3 Context-free (Type 2) grammars

[These grammars differ from the Type 1 in two respects: first, the string x to be rewritten must have length one (this turns out to be only a formal difference since the generative power of Type 1 grammars is unaltered by this restriction); second, and more germane, no context can be used to decide how to rewrite x.]

6.2.1.4 One-sided linear (Type 3) grammars

[These are context-free grammars for which the string y that x is rewritten as must have a very special form: it must be of the form α where α is a string of so-called terminal (or, nonrewritable) symbols, or of the form αY where α is a string of terminals and Y is a single rewritable symbol; the generative power of such grammars is unaltered if the form $Y\alpha$ is used instead of the form αY, but this power is strictly increased if one permits both forms in the same grammar.]

6.2.2 One-sided linear languages

6.2.2.1 Characterization by finite automata (see 6.1.3)

6.2.2.2 Characterization by regular expressions

[This characterization asserts that the Type 3 languages are the smallest family of languages containing the finite sets and closed under a finite number of applications of the operations of union, concatenation and Kleene closure.]

6.2.2.3 Subclasses of the finite-state languages

[Included here are definite languages, noncounting languages, star-height-restricted languages.]

6.2.3 Context-free languages and grammars

6.2.3.1 Characterization by nondeterministic pushdown automata (see 6.1.5.2)

6.2.3.2 Ambiguous grammars and inherently ambiguous languages

[Certain grammars generate some of their strings in more than one way, thus impeding the process of parsing the strings generated; even worse, certain context-free languages are generated only by such ambiguous grammars; i.e., they are _inherently_ ambiguous.]

6.2.3.3 Canonical forms

[Among the most important canonical forms are Chomsky Normal Form which insists that a string x can be rewritten only as a string y of length at most two, and Greibach Normal Form which demands that x be rewritten as a string of the form ay where a is a terminal symbol; any context-free grammar can be replaced by an equivalent one in either of these forms.]

6.2.3.4 Closure properties

[It has proven fruitful in the study of formal languages to determine how a class of languages "behaves" when its languages are operated on in a variety of ways: relevant operations include union, intersection, relative complement, string homomorphism, concatenation, iteration, substitution.]

6.2.3.5 Decision properties

[Basic decision questions include deciding: emptiness, universality, inherent ambiguity, disjointness, acceptability by a _deterministic_ pushdown automaton, acceptability by a finite-state acceptor; see also 6.6.5.]

6.2.3.6 Algebraic characterizations

[Included here are characterization in terms of

operations on languages and characterization in terms of so-called formal power series.]

6.2.4 Parsing and recognition of context-free languages

[Context-free languages are often used for describing the core of the syntax of programming languages; in this role, the problem of efficient parsing is a basic one.]

6.2.4.1 Special forms of grammars

[Included here are LR grammars, LALR grammars, precedence grammars, Greibach Normal Form grammars; each of the foregoing classes of grammars generates either all context-free languages or some subclass thereof; the grammatical classes are recommended by their making the parsing process more efficient in some respect.]

6.2.4.2 General purpose parsing algorithms

[Included here are the backtracking schemes that simulate nondeterministic pushdown automata, Younger's n-time parsing algorithm, and Valiant's reduction of the process of parsing to the problem of transitive closure (thus yielding a time $n^{2.78}$ algorithm).]

6.2.4.3 Complexity of parsing and recognition

[The standard complexity measures are the time and space requirements of the algorithms that perform the desired task; less basic measures of complexity include the number of "tape reversals" and the number of "nondeterministic steps"; models studied include Turing machines, idealized random access machines, idealized list processing machines, iterative arrays of finite-state acceptors, and a number of variants of pushdown automata; see 6.1.5.]

6.2.5 Restricted context-free languages

6.2.5.1 Deterministic context-free languages

[These are the languages accepted (or, recognized) by <u>deterministic</u> pushdown automata.]

6.2.5.2 Linear and metalinear languages

[These subclasses of the context-free languages arise from restrictions on the form of grammatical rules.]

6.2.6 Context-sensitive languages

6.2.6.1 Characterization by linear-bounded automata

[<u>See</u> 6.5.2.1; a problem outstanding for roughly two decades is whether or not the context-sensitive languages are accepted also by the deterministic linear-bounded automata.]

6.2.6.2 Restricted context-sensitive languages

[Included here are: "time-bounded" context-sensitive

grammars for which words generated by the grammar are ignored unless the length of their derivation is bounded above by a given function of the length of the word; "tape-restricted" context-sensitive languages that are accepted by Turing machines whose tape requirements are strictly less than linear in the size of the input (_see_ _also_ 6.5.1, 6.5.2, 6.5.3); and "bounded-context" grammars, including context-sensitive grammars that must produce at least one terminal symbol at each step and grammars that use only terminal context.]

6.2.7 Departures from the Chomsky paradigm

6.2.7.1 Generalized rewriting systems

[Included here are: indexed grammars, grammars with control sets, programmed grammars, macro grammars, parallel rewriting systems, and developmental systems.]

6.2.7.2 Abstract families of languages

[Whereas 6.2.7.1 is concerned with classes of languages that arise by altering the way that grammars generate strings, the current notion considers classes of languages that are generated from given sets using a given repertoire of operations much in the way that finite-state languages (see 6.2.2.2) are generated from the finite sets using the operations of union, concatenation and Kleene closure.]

6.3 Analysis of Programs

[Determination of the properties, meaning and structure of programs.]

6.3.1 Program syntax

[Representation of programs at the level of character strings and tree structures.]

6.3.1.1 Concrete syntax

[Representation of programs as strings; associated with formal language theory and parsing; _see_ _also_ 6.2.]

6.3.1.2 Abstract syntax

[Representation of programs as trees; associated with word algebras.]

6.3.2 Program semantics

[What programs mean; needed for implementation, manuals, verification.]

6.3.2.1 Operational semantics

[Definition of program meaning in terms of the machines they run on or interpreters they run under.]

6.3.2.1.1 SECD machine

[Landin's Stack-Environment-Control-Dump machine for defining meaning.]

6.3.2.1.2 Vienna Definition Language

[A semantic method developed at IBM Vienna; applied most notably to formal definition of PL/I.]

6.3.2.1.3 Lambda-calculus models

[Church's λ-calculus is particularly well suited to supplying operational semantics.]

6.3.2.1.4 Combinatory logic models

[Combinatory logic is a variable-free version of the λ-calculus.]

6.3.2.1.5 Petri nets

[A graph-based operational semantics for parallel control; uses tokens traversing a graph.]

6.3.2.2 Denotational semantics

[Definition of a program as an abstract mathematical object; no explicit mention of an interpreter.]

6.3.2.2.1 Command sequences

[A program is defined as a set of strings of commands; no mention of states.]

6.3.2.2.2 State trajectories

[A program is defined as a set of sequences or trajectories of states; no mention of commands.]

6.3.2.2.3 Binary relations

[A program is a set of initial-final pairs of states.]

6.3.2.2.4 Lattices

[Ongoing computations are considered to be accumulating information; the lattice ordering is interpreted as an ordering on information content.]

6.3.2.2.5 Complete partial orders

[Asymmetric alternative to lattices.]

6.3.2.2.6 Varieties, or equational classes of algebras

[For defining/specifying data types equationally.]

6. THEORY OF COMPUTATION

6.3.2.3 Axiomatic semantics

[Methods for proving properties of programs are sometimes taken as defining the meaning of programs; see also 6.3.4.]

6.3.3 Program constructs

[The building blocks of programs.]

6.3.3.1 Assignment

[A mechanism for storing a datum in a location.]

6.3.3.1.1 Unscoped assignment

[Assignment intended to remain in force until superseded by another assignment; associated with global variables.]

6.3.3.1.1.1 Simple assignment

[Assignment to a variable.]

6.3.3.1.1.2 Subscripted assignment

[Assignment to an array location indexed by integers.]

6.3.3.1.1.3 Record assignment

[Assignment to a record location indexed by field.]

6.3.3.1.2 Scoped assignment

[Assignment associated with a local variable.]

6.3.3.1.2.1 Parameter mechanisms

[Assignment to a formal parameter of a procedure; associated with call by name, value, reference, etc.]

6.3.3.1.2.2 Lexical scoping

[Assignment to variable x local to expression e affects only references to x made explicitly in e.]

6.3.3.1.2.3 Fluid (dynamic) scoping

[Assignment to variable x local to expression e applies to every occurrence of x encountered until current execution of e terminates.]

6.3.3.2 Control (see also 4.1.1.4)

[Determines the commands executed by the computer.]

6.3.3.2.1 Sequential control

[Control constructs appropriate for a single sequential machine.]

6.3.3.2.1.1 Conditional statements

[Permits choice of actions predicated on a test.]

6.3.3.2.1.2 Iteration

[For repeating the same action a number of times.]

6.3.3.2.1.3 Procedure calls

[For naming and later invoking an action.]

6.3.3.2.1.4 Recursion

[The situation that obtains when a procedure calls itself directly or indirectly.]

6.3.3.2.2 Parallel control

[For more than one virtual or real processor.]

6.3.3.2.2.1 Serializable concurrency

[For interleaved command sequences; modelled by the shuffle operator.]

6.3.3.2.2.2 Asynchronous concurrency

[For independent parallel processing; modelled by Petri nets.]

6.3.4 Program logic

[Methods of reasoning about programs; used for manual and automatic verification that programs meet their specifications.]

6.3.4.1 Flowchart logics

[Assertions tag the arcs of flowcharts; graph form of partial correctness.]

6.3.4.2 Partial correctness logics

[Assertions bound from above the possible input/output behavior of a program.]

6.3.4.3 Termination logics

[Assertions bound from below the possible input/output

behavior of a program; dual of partial correctness.]

- 6.3.4.4 Algorithmic logics

    [Program logics closed under Boolean operations and containing usual partial correctness assertions.]

- 6.3.4.5 Modal logics

    [Program logics utilizing time sequences in tense logics.]

6.3.5 Program optimization

[Reorganization of programs to improve efficiency.]

- 6.3.5.1 Flow analysis

    [Extracting global topological properties of programs to aid optimization decisions.]

- 6.3.5.2 Recursion removal

    [Transformation of recursive programs to reduce space attributable to stack use.]

6.3.6 Program schematology

[Study of program properties independent of interpretations of function and predicate symbols.]

- 6.3.6.1 Comparative expressiveness

    [For example, recursive schemas are more expressive than iterative ones.]

- 6.3.6.2 Decision problems

    [For example, the halting problem for one-variable iterative schemes is decidable but, for two-variable schemes, it is undecidable.]

6.4 Computer Models

[Abstract models which capture some of the properties of real computers.]

6.4.1 Turing machines (see also 6.1.5.4)

[Devices with a finite-state memory plus an unbounded, two-way, alterable tape memory.]

- 6.4.1.1 One-tape machines

    [All input, output and computation occur on a single tape.]

    - 6.4.1.1.1 Turing's original model

        [Given the present state and the type symbol currently scanned under the tape's read-write head, the following three actions are uniquely determined: write a symbol on the scanned tape square, shift the tape at most one square, enter a new state.]

6.4.1.1.2 Post's variant

[A machine which cannot both alter the tape symbol being scanned and shift the tape on the same step.]

6.4.1.1.3 Wang's variant

[The machine is programmed in single-address fashion; typical instructions are "shift tape right," "write a 1 on the tape," "go to 5 if the tape symbol being scanned is a 1."]

6.4.1.2 Multitape machines

[More than one tape is allowed; the action of the machine is determined by the present state and a vector of scanned symbols.]

6.4.1.2.1 Ordinary tapes

[The tapes are linear, two-way devices with a single head per tape.]

6.4.1.2.2 Multihead tapes

[Each tape may have several heads, capable of independent motion.]

6.4.1.2.3 Multidimensional tapes

[Each "tape" is a multidimensional array.]

6.4.2 Random-access machines

[Devices having a set of registers, each of which can hold an arbitrary nonnegative integer; program branching can occur by testing the contents of the registers.]

6.4.2.1 Shepherdson-Sturgis machines

[The instruction set enables the machine to: copy from one register to another, increase or decrease the contents of a register by 1, unconditional branch, branch if the content of a register is 0.]

6.4.2.2 Loop machines (see also 6.3.3.2.1)

[The branch instructions are replaced by nested loop control in which the contents of a register will determine the number of times a particular loop is executed.]

6.4.2.3 Variants

6.4.2.3.1 Indirect addressing

6. THEORY OF COMPUTATION

[Registers may be accessed indirectly; this permits a given program to access an unbounded number of registers.]

6.4.2.3.2 Bound on register contents

[Register contents cannot exceed a given limit; alternatively, register contents cannot be increased except by copying.]

6.4.2.3.3 Larger instruction sets

[Instructions to add, subtract, multiply, etc. are permitted.]

6.4.3 Iterative arrays (see also 6.1.5.5)

[An embedding of identical finite-state machines in a regular, possibly infinite lattice, so that each machine is connected to its immediate neighbors in the lattice; also called tessellation automata.]

6.4.3.1 One-dimensional arrays

[Machines are arranged in a linear array, indexed by the integers or the nonnegative integers.]

6.4.3.1.1 Serial I/O at end of array

[Input/output occurs through the machine with index 0, one symbol at a time.]

6.4.3.1.2 Parallel I/O to all elements

[Each machine in the array simultaneously receives one symbol of the input string and emits one output symbol.]

6.4.3.1.3 Common bus

[Each machine in the array has access to a single bus (i.e. communication channel).]

6.4.3.2 Higher dimensional arrays

[The lattice is an array of higher dimension, i.e., machines are indexed by n-tuples of integers.]

6.4.3.3 Pattern reproduction in two-dimensional arrays

[A complex pattern of states at a given time may at a later time appear in more than one place, yielding a kind of self-reproduction.]

6.5 Complexity of Computations

[Complexity is concerned generally with how much effort or resources are required to perform a computation.]

6.5.1 Time complexity

[The number of program steps required for a computation.]

6.5.1.1 Real time

[The computer must read one input symbol per step and produce the outputs without delay.]

6.5.1.2 Linear time

[The total time of a computation is bounded by a linear function of the length (i.e. the number of symbols) of the input.]

6.5.1.3 Polynomial time (see also 6.6.3.1)

[The total time of a computation is bounded by a polynomial function of the length of the input.]

6.5.1.4 Exponential time

[The total time of a computation can grow exponentially with the length of the input.]

6.5.2 Space complexity

[The number of units of storage required for a computation.]

6.5.2.1 Linear space

[The amount of storage required for a computation is bounded by a linear function of the length of the input.]

6.5.2.2 Polynomial space

[The amount of storage required is bounded by a polynomial function of the length of the input.]

6.5.2.3 Exponential space

[The amount of storage required can grow exponentially with the length of the input.]

6.5.2.4 Time vs. space

[Complexity measures combining time and space usage; time-space tradeoffs.]

6.5.3 Other measures of complexity

6.5.3.1 Tape reversals

[The number of times the direction of a storage tape is changed.]

6.5.3.2 Conditional transfers

[The number of conditional branch instructions encountered in a computation.]

6.5.3.3 Nondeterministic steps

[The number of times in a computation the next state

is chosen nondeterministically.]

6.5.3.4 Circuit complexity

[The number of gates required to build a circuit to realize the input/output behavior of a particular computation.]

6.5.4 Machine independent complexity theory

[An abstract treatment of measures of complexity.]

6.5.4.1 Basic axioms

[(i) A complexity measure is defined if and and only if the computation runs to completion; (ii) one can effectively check whether or not a computation has exceeded a given resource bound.]

6.5.4.2 Speed-up properties

[Results giving the existence of functions which are so difficult to compute that, given any algorithm for such a function, there must exist another which computes it much faster.]

6.5.4.3 Honest complexity classes

[Classes of functions having output size similar in magnitude to their computational complexity.]

6.6 Analysis of Algorithms

[Analysis of the properties of problems themselves and the algorithms used to solve them; for a given algorithm interest usually focuses on its average performance and worst possible performance as a function of the inputs to it.]

6.6.1 Complexity of numerical problems

[Various areas of numerical mathematics have been particularly affected by the study of algorithm complexity.]

6.6.1.1 Fast Fourier transforms

[A family of numerical algorithms for computing the discrete Fourier transform.]

6.6.1.2 Evaluation of polynomials

[Given the coefficients of a polynomial, the concern is with how fast it can be evaluated.]

6.6.1.2.1 Horner's rule

[A well-known method for evaluating polynomials by alternating multiplications and additions; known to be optimal when evaluating an arbitrary polynomial at a single point.]

6.6.1.2.2 Preconditioning

[Computing ahead of time certain quantities which depend upon the coefficients of a polynomial but not upon the arguments, with the object of speeding up evaluations of the same polynomial at several different arguments.]

6.6.1.3 Bilinear forms

[Algorithms in which all nonscalar multiplications have as the left argument linear combinations of elements of one input vector and as the right argument linear combinations of elements of a second input vector.]

6.6.1.4 Matrix multiplication

[Given two matrices, A and B, the concern is with how fast their product AB can be computed; if A and B are nxn, the standard method requires n^3 multiplications.]

6.6.1.4.1 Reducing arithmetic operations

[e.g., by a method of Strassen which multiplies 2x2 matrices in only 7 multiplications and which can be applied recursively, or more recent methods which multiply certain matrices using fewer multiplications than Strassen's method.]

6.6.1.4.2 Reducing page faults

[Arranging the pagination of large dense matrices so that as often as possible the desired data is in main memory.]

6.6.1.5 Analytic complexity

[Relates the cost of an algorithm for approximating a real number or real function to the accuracy of the approximation.]

6.6.1.6 Parallel algorithms

[Methods in which much of a desired computation can be done concurrently on multiple arithmetic units or pipeline architectures.]

6.6.2 Complexity of nonnumeric and combinatorial problems

[Problems involving symbol rather than numeric manipulations or which are related to combinatorial mathematics; see 5.2.3.]

6.6.2.1 Searching (see also 7.8.2)

[The measure of complexity is usually the number of data elements accessed compared to the element being searched for.]

6.6.2.2 Insertion/deletion

[The cost of altering a data structure when one or more items are added to or deleted from the file.]

6.6.2.3 Sorting (see also 7.8.1)

[The measure of complexity is usually the number of comparisons made of k keys of records as the file is sorted.]

6.6.2.4 Graph algorithms (see also 5.2.2)

[Methods for traversing, manipulating and analyzing graphs.]

6.6.3 Apparently hard problems

[Problems for which no efficient solution is known but which have not been proven intractable.]

6.6.3.1 Polynomial time reducibility

[A problem is polynomial time reducible to a second problem if any question about the first problem can be answered in polynomial time given sufficient information about the second problem.]

6.6.3.1.1 Cook reducibility

[A question about the first problem can be answered in polynomial time given the ability to have a single question of the second problem answered.]

6.6.3.1.2 Karp reducibility

[A question about the first problem can be answered in polynomial time given the ability to have a polynomial number of questions of the second problem answered.]

6.6.3.2 NP-complete problems

[A class of problems which can be solved in nondeterministic polynomial (NP) time in the sense that some polynomial time sequence of operations will solve the problem on a machine where nondeterministic actions are allowed; any two problems in the class are polynomial time reducible to each other; thus all or none are intractable.]

6.6.3.2.1 Satisfiability of a Boolean expression

[Given a Boolean expression, determine whether one can find a way of assigning "true" to some of its variables and "false" to the other variables so that the entire expression has the value "true."]

6.6.3.2.2 Existence of a k-clique

[Given a graph, determine whether or not there exists a subgraph of size k which is complete (i.e. contains edges joining each pair of vertices).]

6.6.3.2.3 Existence of a Hamiltonian circuit

[Given a graph, determine whether or not there is a path which traverses all of the vertices of the graph, visiting each node exactly once, and returning to the initial vertex.]

6.6.3.2.4 Other equivalent problems

[Such as the traveling salesman, integer linear programming and knapsack problems.]

6.6.4 Inherently hard problems

[Problems which have been proven intractable, i.e., problems where the computation time is at least an exponential function of the input.]

6.6.4.1 Equivalence of regular expressions

[Given two regular expressions (cf. 6.1.3.3) determine whether they define the same language.]

6.6.4.2 Presburger arithmetic

[An arithmetic system on integers with addition but not multiplication for which it is decidable whether or not an arbitrary sentence is a theorem.]

6.6.5 General unsolvability results

[Problems which cannot be solved algorithmically.]

6.6.5.1 Halting problem

[Given a computer program and data, to determine whether or not the program will run to completion.]

6.6.5.2 Post correspondence problem

[Given a set of pairs of strings, to determine whether or not one may choose (with repetitions allowed) a sequence of pairs so that the string formed by concatenating all of the left members of the sequence equals the string formed by concatenating the right members of the sequence.]

6.6.5.3 Other equivalent unsolvable problems

[Problems which can be effectively enumerated, i.e., one can detect when the answer is "yes," but which are undecidable in the sense that, if answers were provided by an "oracle," the halting problem could be decided.]

6.6.5.4 Higher degrees of unsolvability

[Problems which cannot even be effectively enumerated, e.g., to determine for a given program whether or not it runs to completion on all possible inputs.]

PART IV. ANNOTATED TAXONOMY TREE
7. METHODOLOGIES

PART IV. ANNOTATED TAXONOMY TREE
7. METHODOLOGIES
7.1 ALGEBRAIC MANIPULATION

7.1 Algebraic Manipulation

[Refers to the body of algorithms and systems which deal with the manipulation of mathematical expressions or formulas in their natural, symbolic form rather than just the manipulation of numbers.]

7.1.1 Expressions and their representation

[The types of mathematical expressions that can be manipulated and the ways in which they are stored in a computer.]

7.1.1.1 Data structures

[The underlying structures, such as lists and strings, used to represent expressions to be manipulated.]

7.1.1.2 Types of expressions

7.1.1.2.1 Polynomials

[Expressions of the form $a_n x^n + a_{n-1} x^{n-1} + \ldots + a_1 x + a_0$ where the coefficients a may involve variables other than x.]

7.1.1.2.2 Rational functions

[Ratios of polynomials.]

7.1.1.2.3 Algebraic functions

[Functions which are solutions of polynomial equations having coefficients which are also polynomials; for example, the square root function.]

7.1.1.2.4 Transcendental functions

[Functions which are not algebraic; includes elementary functions such as exponentials, logarithms, and trigonometric functions.]

7.1.1.2.5 Power series

[Includes representation of truncated Taylor and Laurent series.]

7.1.1.2.6 Composite objects

[Includes representations for matrices and sets.]

7.1.1.2.7 Other

[Includes specialized representations for expressions occurring in general relativity or astronomy, for example.]

7.1.1.3 Representations of expressions

7.1.1.3.1 Polynomial representations (see 7.1.1.2.1)

7.1.1.3.1.1 Distributed

[The usual way of representing polynomials as a sum of terms.]

####### 7.1.1.3.1.2 Recursive

[In which, for example, a polynomial in two variables, x, y, would be represented as a polynomial in x with coefficients which are polynomials in y.]

####### 7.1.1.3.1.3 Factored

[A product of sums format, e.g., expressing a polynomial as a product of two cubics, neither of which can be further factored over the integers.]

7.1.1.3.2 General representations

[Representations which are more general than polynomial representations and permit the representation of nonrational functions, such as exp(x) or sin(x), with relative ease.]

7.1.1.4 Numerical coefficient domains

[The values which the constant coefficients of an expression (e.g. a polynomial) are allowed to assume.]

7.1.1.4.1 Infinite precision integers

[Integers which are not limited in size to a fixed number of digits.]

7.1.1.4.2 Rational numbers

[Ratios of integers.]

7.1.1.4.3 Algebraic numbers

[Numbers which are the zeros of polynomials with rational coefficients; includes Gaussian integers and irrationals such as the square root of 2.]

7.1.1.4.4 Floating-point numbers (see 5.2.1.2)

7.1.1.4.5 Variable precision floating-point numbers

[Permits the setting of the precision (i.e. the number of digits in the mantissa) of floating-point numbers.]

7.1.2 Algorithms

[Procedures that manipulate algebraic expressions.]

7.1.2.1 Operations on polynomials, rational functions and

7.1 ALGEBRAIC MANIPULATION

coefficient domains

7.1.2.1.1 Basic arithmetic operations

[Addition, subtraction, multiplication, and division.]

7.1.2.1.2 Greatest common divisor algorithm

[Extension of Euclid's algorithm to polynomials; needed, for example, to eliminate common factors in rational functions.]

7.1.2.1.3 Factorization

[The process of converting, for example, a polynomial in recursive representation (see 7.1.1.3.1.2) to factored representation (see 7.1.1.3.1.3).]

7.1.2.1.4 Determinants and solutions of linear equations

[The elements or coefficients are generally symbolic rather than numerical.]

7.1.2.1.5 Resultants

[A function of two polynomials that is used to eliminate a variable common to them.]

7.1.2.1.6 Other

[Such as partial fraction expansions and solutions of systems of polynomial equations.]

7.1.2.2 Simplification

[Converting an expression into an equivalent one which is in a more useful form, such as one having fewer terms.]

7.1.2.2.1 Canonical forms

[Standard forms to which classes of expressions can be reduced such as the fully distributed representation for polynomials.]

7.1.2.2.2 Algebraic independence

[Certain simplification algorithms can guarantee that the nonrational functions in an expression, such as exp(x) and log(x), have no polynomial relation among them.]

7.1.2.2.3 Pattern matching and rule systems

[The capability of using additional rules, such as $\sin^2(A)+\cos^2(A)=1$, in order to simplify an expression.]

7. METHODOLOGIES

7.1.2.2.4 Decidability results

[Such as the inability to simplify classes of expressions involving the absolute value function.]

7.1.2.2.5 Other

[Such as approaches to generating identities to special functions from their definition as differential equations.]

7.1.2.3 Differentiation

[The familiar operation used in the calculus.]

7.1.2.4 Limits

[The familiar operation used in the calculus.]

7.1.2.5 Integration

[The familiar operation used in the calculus, with the goal of obtaining a closed-form solution using common techniques (e.g. partial fractions) as well as other highly sophisticated techniques.]

7.1.2.5.1 Indefinite integration

[Where no limits are specified.]

7.1.2.5.1.1 Rational functions (see 7.1.1.2.2)

7.1.2.5.1.2 Exponential and logarithmic functions

[Which include all trigonometric functions as a special case.]

7.1.2.5.1.3 Algebraic function extensions (see 7.1.1.2.3)

7.1.2.5.1.4 Special functions

[Defined by differential equations.]

7.1.2.5.2 Definite integration

[Integration with specified limits, such as 0 and infinity; for the latter, classical techniques such as residue calculations are used.]

7.1.2.5.3 Ordinary differential equations

[Reduction of such equations to indefinite integrals and then solution of these, whenever possible.]

7.1.2.6 Summation

[Algorithms for obtaining a closed form formula for the sum of a rational function, for example, from 0 to n or 0 to infinity.]

7.1.2.7 Analysis of algorithms

[Analysis of the behavior of an algorithm in terms of space and time under various inputs.]

7.1.2.7.1 Worst-case analysis

[An analysis which shows the behavior of the algorithm in the worst (e.g. slowest) case; often this analysis makes additional assumptions about the inputs such as whether they are dense (no missing terms) or sparse.]

7.1.2.7.2 "Fast" algorithms

[A class of algorithms which often depend on having inputs of relatively large size.]

7.1.2.7.3 Exact analysis

[Formulas which predict the exact cost of an algorithm as a function of various parameters (such as the number of variables in the input).]

7.1.2.7.4 Probabilistic analysis

[Algorithms which yield correct answers, but whose computing time analysis depends on a probabilistic analysis.]

7.1.2.7.5 Average computing time analysis

[An analysis which determines the cost of an algorithm on the average over all possible inputs.]

7.1.2.7.6 Other

[Includes, for example, reducibility results which show how one algorithm permits one to obtain the results to a different algorithm with relatively little additional work.]

7.1.2.8 Other

[Includes specialized algorithms for applications in celestial mechanics and high energy physics.]

7.1.3 Languages and systems

[Issues related to the language used in and the design of an algebraic manipulation system.]

7.1.3.1 Type of system

[Differentiates between relatively general purpose

systems and more specialized ones for solving problems, for example, in astronomy or general relativity.]

- 7.1.3.2 Mode of operation

  - 7.1.3.2.1 Interactive

    [Widely used in algebraic manipulation systems since the user often does not know in advance exactly how to proceed from one step to the next; see also 2.2.3.]

  - 7.1.3.2.2 Batch (see 2.2.1)

- 7.1.3.3 Languages

  [Systems which are not simply a collection of subroutines but present a language to use while interacting with them.]

  - 7.1.3.3.1 Extension of existing languages

    [The addition of algebraic manipulation facilities to existing languages such as Fortran.]

  - 7.1.3.3.2 Procedural languages

    [Languages with nontrivial variations in syntax and semantics from existing languages, such as the language for MACSYMA; Lisp is often the implementation language in these cases.]

  - 7.1.3.3.3 Nonprocedural languages (see also 4.1.1.3.7)

    [Languages which allow the user to specify less detail than is currently common in procedural languages; often based on Markov algorithms (e.g. SCRATCHPAD).]

- 7.1.3.4 Evaluation of expressions

  [Methods used in algebraic manipulation systems by which various types of mathematical expressions are symbolically evaluated.]

  - 7.1.3.4.1 Evaluation by substitution

    [To evaluate $f(x)$ at $x=b$, substitute b for x in the expression $f(x)$.]

  - 7.1.3.4.2 Markov algorithms

    [Evaluation by applying a set of rules (often called productions) until no further change occurs.]

  - 7.1.3.4.3 Lambda-calculus evaluation (see 6.3.2.1.3)

- 7.1.3.5 Input/output facilities

7.1 ALGEBRAIC MANIPULATION

7.1.3.5.1 Automatic two-dimensional display of expressions

[Methods for generating textbook-like style display of expressions.]

7.1.3.5.2 Output formatting

[Facilities for controlling the format of the output which have no effect on the internal representation of an expression.]

7.1.3.5.3 Handwritten input of expressions

[Techniques for parsing handwritten expressions on tablet-like devices.]

7.1.3.6 Interface to numerical routines

[Includes techniques for transforming expressions into a form that is accessible by numerical routines.]

7.1.3.7 Specialized processors

[Techniques for specializing an existing processor through firmware or the design of a new processor especially suited for algebraic processing.]

7.1.3.8 Specialized consoles

[Includes special display keyboards and other input or output devices especially suited for algebraic manipulation.]

7.1.4 Applications

[Areas in which nontrivial applications of algebraic manipulation have been made.]

7.1.4.1 Physics

7.1.4.1.1 General relativity

[Calculation of tensors, for example.]

7.1.4.1.2 High energy

[Solution of Feynman diagrams, for example.]

7.1.4.1.3 Plasma physics

[Includes perturbation techniques applied to basic partial differential equations in plasma physics.]

7.1.4.1.4 Other

[Includes applications in optics.]

7.1.4.2 Mathematics

7.1.4.2.1 Group theory

[Includes work on specialized systems and algorithms for the analysis of large finite groups.]

7.1.4.2.2 Hydrodynamics

[Perturbation techniques applied to problems in hydrodynamics.]

7.1.4.2.3 Other

[Includes application in algebraic geometry, algebraic topology, combinatorics, logic, number theory, and statistics.]

7.1.4.3 Computer science

7.1.4.3.1 Analysis of algorithms

[Includes uses of algebraic manipulation in deriving the formulas needed in an analysis of an algorithm.]

7.1.4.3.2 Program verification

[Includes uses of symbolic techniques in proving the correctness of programs.]

7.1.4.3.3 Numerical analysis

7.1.4.3.3.1 Error analysis

[Includes use of algebraic manipulation techniques in automatic error analysis.]

7.1.4.3.3.2 Partial differential equations

[Includes generation of high level language programs which have been optimized to solve a particular partial differential equations using finite element or finite difference techniques.]

7.1.4.3.3.3 Other

[Includes symbolic front-ends to libraries of numerical procedures.]

7.1.4.4 Astronomy

[Includes special purpose systems in celestial mechanics.]

7.1.4.5 Engineering

[Includes applications in control theory and signal analysis.]

7.1.4.6 Education

[Includes applications to mathematics or science education at the high school and college level.]

7.1.4.7 Other

[Such as medical science and meteorology.]

PART IV. ANNOTATED TAXONOMY TREE
7. METHODOLOGIES
7.2 ARTIFICIAL INTELLIGENCE

7.2 Artificial Intelligence

[The study of how to make computers exhibit behavior that is considered "intelligent" when observed in humans; includes the study of how to make computers perform tasks that, until recently, only humans could perform ("Machine Intelligence") and the construction and study of computer executable models of human intelligent behavior ("Simulation of Human Cognitive Behavior").]

7.2.1 Automatic programming (see also 4.1.1.1.4)

[The study of how to construct programs which aid humans in the development of other programs.]

7.2.1.1 Program construction

[Given a statement of a problem in a non-programming language, the program constructs a program to solve the problem.]

7.2.1.2 Program verification

[Given a specification of a problem and a purported program for its solution, the AI program proves whether or not the given program solves the problem.]

7.2.1.3 Programming assistants

[The program provides intelligent aids to a human program developer such as understanding of natural language comments written by the human, automatic debugging, locating and explaining useful subroutines.]

7.2.2 Deduction

[Study of mechanisms for performing deductive inference: manipulating general rules or applying general rules to specific instances; includes theorem proving, deductive question answering, inference methods for planning, problem solving, etc.]

7.2.2.1 Direct methods

[Methods based on rules of inference such as universal and existential instantiation, modus ponens, cut, etc., and on formal systems such as those of Hilbert, Gentzen, Fitch ("natural deduction"), etc.]

7.2.2.2 Refutation methods

[Methods based on proving the inconsistency of the negation of whatever is to be shown with the set of axioms; for example, methods based on the resolution principle.]

7.2.2.3 Informal methods

[Methods employing "common sense" rules of inference or methods of reasoning.]

7.2.2.4 Non-standard logics

[Application of logics other than classical two-valued

logic such as fuzzy logics, higher order logics and modal logics.]

7.2.3 Learning

[Study of the design of programs that behave differently under repeated instances of the same input due to modification of the program or its parameters during each execution.]

7.2.3.1 Concept formation

[Methods involving the formation of models that are then used to decide if some input belongs to some class; includes models in the form of networks, decision trees, etc.]

7.2.3.2 Grammatical inference

[Methods involving the construction of grammars which can then be used for generating elements of a class as well as for deciding if an element is in a class through the use of parsing techniques.]

7.2.3.3 Inductive inference

[Methods involving the formation of general rules which can then be used by methods of deductive inference.]

7.2.3.4 Parameter weighting

[Methods involving the adjustment of coefficients of evaluation or decision functions.]

7.2.4 Natural language processing

[Study of how to design procedures to process natural language as part of programs that use natural language intelligently, as opposed to treating natural language as meaningless symbol strings.]

7.2.4.1 Speech understanding (see also 1.4.1.7)

[Attention is paid to the special problems resulting from the use of acoustic waveforms as input.]

7.2.4.2 Parsing (see also 6.2.4 and 7.3.1.4)

[Translation of natural language strings into more useful data structures.]

7.2.4.3 Discourse analysis

[Attention is paid to the special problems of multi-sentence texts such as articles, stories, dialogues, etc.]

7.2.4.4 Sentence generation

[Generation of natural language sentences from other representations.]

7.2.4.5 Discourse generation

[Generation of multi-sentence texts, paying attention to coherence, development, etc.]

7.2.4.6 Speech production

[Mechanical generation of human speech; includes problems of the production of appropriate stress patterns.]

7.2.5 Problem solving

7.2.5.1 Game playing

[Use of games such as chess to study problem solving in a competitive environment.]

7.2.5.2 Heuristic search

[Study of methods of searching large problem spaces guided by some knowledge about the problem, but where that knowledge is insufficient to guarantee that no false paths will be searched.]

7.2.5.3 Planning

[Study of methods of constructing programs that construct plans to be used, possibly with modification, in later problem solving activities.]

7.2.6 Representation of knowledge (see also 3.1.2)

[Study of formalisms and data structures for the representation of knowledge to be used for natural language understanding, problem solving, scene understanding, etc.]

7.2.6.1 Network representations (see also 7.3.1.4.6)

[Representations based on directed graphs, such as relational graphs, semantic networks, conceptual networks, etc.]

7.2.6.2 Predicate calculus representations

[Representations based on the standard syntax of symbolic logic.]

7.2.6.3 Procedural representations

[Using procedures written in a high level programming language to represent knowledge.]

7.2.7 Robotics

[Study of machines that interact with their (real or simulated) environment by moving around in it and/or manipulating it, and do this in what could be considered an intelligent manner.]

7.2.7.1 Exploration robots

[The machine moves about in its environment, "senses" its environment, avoids crashing into obstacles or falling off precipices, etc.]

7. METHODOLOGIES

7.2.7.2 Manipulation robots

[The machine manipulates its environment or objects (e.g. blocks) in its environment, using both sensors and effectors.]

7.2.8 Vision

[Study of the design of programs that process, categorize and understand visual information.]

7.2.8.1 Two-dimensional objects and line drawings (see also 7.6)

[For example, understanding line drawings of blocks scenes.]

7.2.8.2 Real-world objects and scenes (see also 7.5.8)

[For example, photographs of street scenes, aerial photographs, x-rays of organs.]

7.2.8.3 Motion

[For example, successive frames of a motion picture.]

7.2.9 Specialized systems

[Study of the design of programs that behave intelligently in a specific domain; emphasis on formulation and use of expert knowledge of the domain.]

7.2.9.1 Chemistry systems

[For example, inferring molecular structures from mass spectrometry data.]

7.2.9.2 Intelligent computer-assisted instruction (see also 8.3)

[For example, mixed initiative tutorial programs.]

7.2.9.3 Medical systems

[For example, computer-assisted diagnosis.]

7.2.9.4 Other

[For example, restricted domain natural language information retrieval.]

7.2.10 Software

[Study of programming techniques for creating "intelligent" programs that are useful for more than one AI area, but not necessarily yet ready for general programming applications.]

7.2.10.1 Control structures (see also 4.1.1.4.1)

7.2.10.1.1 Hierarchical

####### 7.2.10.1.1.1 Recursion (see 4.1.2.3)

####### 7.2.10.1.1.2 Backtracking

[A technique, useful in AI and other areas, in which possible solution paths are explored until refuted after which the steps are retraced and side effects undone until a point is reached where another possible path is encountered.]

7.2.10.1.2 Heterarchical

7.2.10.1.2.1 Data-directed

[What is to be done next depends on the current contents of a changing data base.]

7.2.10.1.2.2 Communicating coroutines

[Independent processes send messages to and schedule each other.]

7.2.10.2 Languages

[The study and design of languages that incorporate useful AI techniques as basic routines, data structures and/or control structures; examples are the languages PLANNER, SAIL and STRIPS.]

PART IV. ANNOTATED TAXONOMY TREE
7. METHODOLOGIES
7.3 INFORMATION STORAGE AND RETRIEVAL

7.3 Information Storage and Retrieval

[Is concerned with content analysis and assignment of content identifiers to information items, file organization, query formulations, search and retrieval methods, and retrieval system evaluation.]

7.3.1 Information analysis

[Specification of content identifiers, and the choice of techniques useful in content analysis.]

7.3.1.1 Indexing methods

[Methods concerned with the generation of index terms to be attached to information items for content representation.]

7.3.1.1.1 Term extraction

[Extraction of important terms from text to be used as index terms.]

7.3.1.1.2 Term weighting

[Assignment of weights to index terms which represent the relative importance of the terms.]

7.3.1.1.3 Word truncation

[Truncation of words to stems so that one stem can represent a group of words in the index (e.g. elimination of prefixes and/or suffixes).]

7.3.1.1.4 Thesaurus methods

[The use of synonyms and related word groupings.]

7.3.1.1.5 Phrase formation

[Extraction of important phrases which then become part of the index.]

7.3.1.1.6 Term associations

[Correlated or associated term pairs become content identifiers in the index.]

7.3.1.1.7 Citation indexing

[Chains of documents which cite each other in their bibliographies can be used for content representation.]

7.3.1.1.8 KWIC indexing (see 7.3.1.3.4)

7.3.1.1.9 Indexing language construction

[Construction of a systematic arrangement of index terms by one or more of the previously cited techniques.]

7.3.1.2 Abstracting procedures

[The automatic construction of document abstracts from the full text.]

7.3.1.2.1 Sentence extraction

[Extraction of key sentences from the text to be used in the abstract.]

7.3.1.2.2 Sentence scoring

[Assigning weights to each sentence in its order of importance for content representation.]

7.3.1.2.3 Coherence criteria

[Methods for determining relationships between different sentences used in generating abstracts.]

7.3.1.3 Dictionaries and thesauruses

[The use of various types of dictionaries and word lists for content analysis purposes.]

7.3.1.3.1 Word stem dictionary

[Lists word stems together with all the complete words represented by each stem.]

7.3.1.3.2 Synonym dictionary (thesaurus) (see also 7.3.1.1.4)

[An organized grouping of synonyms and related words.]

7.3.1.3.3 Phrase dictionary

[The entries are word phrases.]

7.3.1.3.4 Keyword-in-context (KWIC) list

[An index formed of individual, significant words is listed in alphabetic order in the context of adjoining words in documents in which they occur.]

7.3.1.3.5 Concept hierarchy

[The content identifiers representing concepts are organized hierarchically (e.g. algebra as a subdivision of mathematics).]

7.3.1.3.6 Concordance utilization

[An alphabetic list of words and phrases appearing in a document with an indication of the place where these words and phrases appear; see also 7.3.1.3.4.]

7.3.1.4 Linguistic processing

[The use of syntactic and semantic techniques as a part of the content analysis and indexing tasks.]

7.3.1.4.1 Phrase structure grammar

[Consists of a set of juxtaposed or nested phrases from sentences.]

7.3.1.4.2 String grammar

[A string of characters generated from text by certain grammatical rules.]

7.3.1.4.3 Augmented transition network

[A representation of the syntactic elements in a language by the nodes of a network whose paths indicate the transition from one component to another.]

7.3.1.4.4 Transformational grammar (see also 6.2.1)

[A grammar consisting of a series of transformational rules reflecting different grammatical uses to express the same essential idea.]

7.3.1.4.5 Case grammar

[A grammar based on semantic features (such as "actor" or "agent" etc.) attached to the elements of a language.]

7.3.1.4.6 Semantic network (see also 7.2.6.1)

[Semantic relations between words are represented by appropriate connections in a network structure.]

7.3.1.4.7 Combined syntactic-semantic systems

[Linguistic approaches combining, for example, those of 7.3.1.4.3 and 7.3.1.4.6.]

7.3.2 Information storage

[The choice of appropriate storage organization for the information records.]

7.3.2.1 Storage organization

7.3.2.1.1 Sequential file organization

[Records stored sequentially with no other organization imposed and no auxiliary index.]

7.3.2.1.2 Chained (multilist) files

[Groups of items, often sharing a common keyword, are stored as lists with pointers

7. METHODOLOGIES

("chains") to facilitate both search and insertion and deletion.]

7.3.2.1.3 Inverted files

[An auxiliary index is maintained for each keyword which contains pointers to all records in which that keyword appears, thus allowing immediate access to these records.]

7.3.2.1.4 Scatter storage files

[Hashing techniques (see 7.8.3.4) are used to store and gain access to the file.]

7.3.2.1.5 Clustered files

[Records are grouped so that items which are sufficiently similar are stored in a common group or cluster.]

7.3.2.2 Record classification

[Methods for automatically constructing groups or classes of records for all items that are sufficiently similar.]

7.3.2.2.1 Hierarchical grouping methods

[Record classes are grouped hierarchically in such a way that larger, more heterogeneous groups of records are broken down into smaller groups containing more and more homogeneous items.]

7.3.2.2.2 Iterative partitioning

[An initial, rough classification is refined by shifting items from one class to another until a satisfactory classification is obtained.]

7.3.2.2.3 Fast single-pass clustering

[Special methods for grouping records on a single pass over the items.]

7.3.2.3 Collection control

[Methods used to control the contents and size of a collection of records and to keep the file up-to-date.]

7.3.2.3.1 Collection growth

[Methods used to incorporate new incoming items into a file.]

7.3.2.3.2 Collection retirement

[Methods used to delete older or unwanted items from a file.]

7.3.2.3.3 Cooperative collection development

[Methods used to pool the resources of several organizations, thus permitting a shared use of several collections.]

7.3.2.3.4 Bibliometrics

[Describes quantitative techniques to measure file sizes and collection properties; used in planning for improvements in system operations and collection development.]

7.3.3 Information search and retrieval

7.3.3.1 Query formulation

[Procedures for formulating and structuring the queries to be used in interrogating a file.]

7.3.3.1.1 Controlled vocabulary

[Query terms are chosen from a limited set of terms contained in a previously approved list.]

7.3.3.1.2 Natural language vocabulary

[Queries are formulated in normal English phrases or sentences; see also 7.2.4.]

7.3.3.1.3 Boolean query formulation

[Simple queries are compounded using Boolean connectives; see also 6.1.1.1.]

7.3.3.1.4 Term vector formulation

[Queries are formulated as vectors of keywords.]

7.3.3.2 Search process (see 7.8.2 and 3.2.3)

7.3.3.3 Retrieval process

[Concerned with how decisions are made on whether or not to retrieve a particular record.]

7.3.3.3.1 Association and correlation coefficients

[Concerned with similarity measures between information items and queries.]

7.3.3.3.2 Query-record similarity

[Measures similarity between a query and a particular record.]

7.3.3.3.3 Record-record similarity

[Measures closeness of a record to another record already retrieved.]

7.3.3.4 Interactive query processing

7. METHODOLOGIES

[Methods permitting improvements in query formulation and record identification through user-system interaction.]

7.3.3.4.1 Vocabulary display

[User reacts to vocabulary words presented on a display screen.]

7.3.3.4.2 Document display

[Whole or parts of previously retrieved documents are displayed to the user.]

7.3.3.4.3 Multistep query refinement

[User reformulates queries based on displayed vocabulary and on previously retrieved documents.]

7.3.3.4.4 Automatic feedback process

[Query refinement is based on information supplied by the user concerning relevance or usefulness of previously retrieved items.]

7.3.4 Retrieval evaluation

[Evaluation of methods of retrieval in terms of search time and cost, and retrieval effectiveness in terms of relevant items retrieved and nonrelevant items rejected.]

7.3.4.1 Efficiency

7.3.4.1.1 Cost-based measures

7.3.4.1.2 Time-based measures

7.3.4.2 Effectiveness

7.3.4.2.1 Recall

[The proportion of relevant items retrieved.]

7.3.4.2.2 Precision

[The proportion of retrieved items which are relevant.]

7.3.4.2.3 Single-valued measures

[Measures expressing retrieval effectiveness using a single expression.]

7.3.5 Retrieval models

[The generation and use of abstract models of the retrieval process.]

7.3.5.1 Set-theoretic models

[The retrieval process is treated as a mapping between partially ordered sets of query terms and partially ordered sets of stored records.]

7.3.5.2 Probabilistic models

[Records are retrieved on the basis of probabilistic parameters expressing the likelihood of relevance given the occurrence of certain terms in certain records.]

7.3.5.3 Decision theoretic models

[These express the loss caused by the retrieval of unwanted items and the loss caused by the rejection of wanted items using a minimized loss function.]

7.3.5.4 Information theoretic models

[The retrieval system is modeled as a communications system.]

7.3.6 Information systems

[A variety of information retrieval and text processing systems.]

7.3.6.1 Storage and retrieval systems

7.3.6.1.1 Reference retrieval systems

[Systems in which references (i.e. documents) are retrieved in answer to submitted queries.]

7.3.6.1.2 Question-answering (fact retrieval) systems

[Systems in which direct answers are supplied to submitted queries.]

7.3.6.1.3 Current awareness (SDI) systems

[Systems in which user profiles expressing user interests are continuously compared against new, incoming items in order to keep professionals abreast of current literature in their fields.]

7.3.6.1.4 Structured database systems (see also 7.4)

[Systems based on the storage of a formatted, business-type database.]

7.3.6.1.5 Integrated information systems

[Information from a variety of domains is integrated into a single system such as a management information system.]

7.3.6.1.6 Information network

[The location of the files in the system may be dispersed over a network.]

7. METHODOLOGIES

7.3.6.2 Auxiliary information systems

[The primary purpose of the system is other than the storage or retrieval of the data in it.]

7.3.6.2.1 Source data automation

[Data is captured and entered into the information system automatically.]

7.3.6.2.2 Text editing (see 8.4.1)

7.3.6.2.3 Automated publication process

[A combined system for text editing, manuscript preparation and automatic typesetting.]

7.3.6.2.4 Automated office systems (see also 8.4.3)

[A generalized information system for servicing the needs of a business office.]

PART IV. ANNOTATED TAXONOMY TREE
7. METHODOLOGIES
7.4 DATABASE MANAGEMENT SYSTEMS

7.4 Database Management Systems

[Software designed to facilitate use and control of large information files maintained on auxiliary storage devices; includes availability of data description and data manipulation languages which permit insertion, deletion and modification of data, provide for data and system integrity and allow security control and data administration.]

7.4.1 Theory

[Refers to the precise, formal, mathematically-oriented study of the operations on and properties of data.]

7.4.1.1 Data semantics

[Is concerned with the meaning of the data (what is generally defined as "information"), the interpretation of data, including valid operations, (e.g. arithmetic), integrity constraints, and validity of various insertion/deletion/modification rules.]

7.4.1.2 Normalization

[A step-by-step process for replacing complex relationships among data with tables such that at every row and column position there is precisely one value (first normal form), or the further reduction of single tables into two or more smaller ones in order to eliminate undesirable intercolumn dependencies (second and third normal forms).]

7.4.1.3 Conceptual schema

[A term which has come to mean the highest level description of the data of an organization; normally it is a model of the actual information flow to, from and in the organization.]

7.4.1.4 Data model theory

[Refers to the comparative study of data models (see 7.4.2.1), with the aim of finding a unifying framework.]

7.4.1.5 Formal mathematical descriptions

[Precise descriptions of data models based on set theory, extended set theory, graph theory, formal logic, or some well defined algebra whose goal is to produce proofs of equivalence of data models or verifications of correctness of representations.]

7.4.2 Logical design

[By which the real environment is abstracted into a set of explicit data descriptions which are complete and consistent for a given environment.]

7.4.2.1 Data models

[The permissible logical data structures and valid/meaningful operations on the data structures.]

7.4.2.1.1 Hierarchical

[A model in which each node of a hierarchical definition tree represents a record type and where the parent-child relationships in the tree represent dependencies between record types.]

7.4.2.1.2 Network

[A model in which nodes represent records of a particular type and the arrows represent relationships between the various record types.]

7.4.2.1.3 Relational

[A formal model which represents relationships among attributes of data types in the form of tables in which each table defines the relation between the attributes of two data types.]

7.4.2.1.4 Set-theoretic

[A model in which relations among data are expressed as sets and the sets are manipulated by set operations.]

7.4.2.2 Syntactic representation

[Of the language used for describing a data model to the computer.]

7.4.2.3 Semantic representation

[The method of describing the data semantics, usually in the form of what operations on the data are valid.]

7.4.2.4 Data dictionary/directory

[A complete list of the data names, a description of their meaning, permissible range of values, security constraints, authority, source, etc. used in a data base.]

7.4.2.5 Data quality and integrity

[Refers to design techniques for improving the quality or accuracy of individual data items and the relationships among data items in order, for example, to protect against tampering as well as to replace damaged or violated data.]

7.4.2.6 Data independence

[Measures how well, from an application programs point of view, the processing of data is insulated from changes in the organization of the data.]

7.4.3 Accessing methods

[Methods whose function is to present the database management system with an interface to stored records.]

- **7.4.3.1 Index organization and searching techniques**

    [Methods, usually based on multilevel structures, to facilitate the search for record occurrences.]

- **7.4.3.2 Indexed access**

    [Record occurrences are accessed through the use of one or more indexes.]

- **7.4.3.3 Direct access**

    [The address of a record is computed as a function of a primary key value that appears in the record (in the degenerate case the primary key value may actually be the address itself).]

- **7.4.3.4 Link and selector access**

    [A technique used in the implementation of relational databases that provides an access path to a subset of tuples in a stored relation.]

7.4.4 Language interfaces

[How humans communicate with the DBMS and related software.]

- **7.4.4.1 Data description languages**

    [Languages used to describe the logical organization of the data in a database.]

    - **7.4.4.1.1 Schema**

        [The global description of the structure of the data accessible in a database.]

    - **7.4.4.1.2 Subschema**

        [A description of a subset or portion of the schema in a form convenient for particular applications.]

    - **7.4.4.1.3 Stored data**

        [A description of the format of the data as it exists on the physical storage device(s).]

    - **7.4.4.1.4 Mapping between levels**

        [Refers to the conversion of data representations from one data description to another.]

- **7.4.4.2 Data manipulation languages**

    [Languages which specify the allowable interactions between a database management system and its users through a set of operations to access and manipulate the data in the database.]

7. METHODOLOGIES

7.4.4.2.1 Self-contained

[A complete language for obtaining data from and manipulating data in a database.]

7.4.4.2.2 Host-embedded

[A language to provide the desired data manipulation operations which is incomplete in itself and which must, therefore, be embedded in a higher level language such as Cobol or PL/I.]

7.4.4.3 Query languages

[Languages designed to facilitate interrogation of the database by humans; see also 7.3.3.1.]

7.4.4.3.1 Keyword

[In which natural language keywords, such as SELECT, FIND, WHERE or INSERT, are used as the basis for describing queries.]

7.4.4.3.2 Two-dimensional

[In which the location in a plane of a shape, character or keyword is used as the basis for describing queries.]

7.4.4.3.3 Specification

[In which the result desired rather than the procedure for arriving at it is described (e.g. Print SCORE where SCORE >10).]

7.4.4.3.4 Procedural

[In which the process for arriving at a result is described rather than specifying the result (e.g. take first column and delete items greater than 60).]

7.4.4.3.5 Menu

[In which the user chooses from a list of alternative commands or data names.]

7.4.4.3.6 Natural language (see also 7.2.4)

[In which the user describes a query in ordinary English or other natural language form.]

7.4.4.4 Protection specification languages

[Languages for communicating data security policies and protection requirements to the database system.]

7.4.4.5 Implementation of language interfaces

[How data manipulation languages and related software

7.4 DATABASE MANAGEMENT SYSTEMS

are implemented to allow humans to communicate with a database management system.]

7.4.4.5.1 Integrating database constructs in high level languages

[Includes provision in high level languages of type definition facilities capable of modeling real world entities and of the capability to reference and manipulate objects in a database.]

7.4.4.5.2 Embedding data manipulation languages in high level languages

[Providing a user interface to database management systems by embedding a data manipulation language in a high level language.]

7.4.4.5.3 Linking programming languages to database management systems

[Setting up interfaces in the form of I/O and/or communications areas and providing a data manipulation language in a programming language.]

7.4.5 Physical design

[Refers to physical organization on storage devices of the database where the central concern is compactness, speed etc.]

7.4.5.1 Software architecture

[How the procedures of the database management systems are structured and integrated with other procedures into well defined program modules.]

7.4.5.2 File management

[Techniques for organizing and searching for data on physical storage devices (see also 7.8.2 and 3.2.3).]

7.4.5.2.1 Access strategies

[Techniques for locating a particular data item.]

7.4.5.2.2 Storage strategies

[Techniques for representation and placement of data on physical storage devices; includes coding, compression, the use of pointers and storage allocation techniques.]

7.4.5.2.3 Buffer management

[Techniques for manipulating buffers for records in main storage (areas to which records are read prior to processing or from which they are written to auxiliary storage after processing) so as to improve efficiency or speed.]

7.4.5.3 Concurrency

[Techniques for permitting multiple users to access and modify a database simultaneously.]

7.4.5.4 Dynamic restructuring

[Techniques for physically reorganizing or logically restructuring a database after, for example, many insertions and deletions have been made.]

7.4.5.5 Data translation

[Conversion of a database from one storage strategy to another which must be done if the computer operating system, database management system or application programs are changed.]

7.4.5.6 Modeling and performance evaluation

[Computer simulation, analytic models, software probes and hardware monitors are used to predict and assess performance, and to determine the effect of logical or physical database structure changes.]

7.4.5.7 Data integrity

[Assuring the accuracy of data and the prevention of inadvertent or deliberate destruction or modification of data items.]

7.4.5.8 Database recovery

[Restoring the database and restarting operations on it after hardware or software failure.]

7.4.5.9 Security protection

[Protection of data in the database against unauthorized disclosure, alteration or destruction.]

7.4.5.10 Hardware and machine architecture

[Organization of computer and related hardware to support database management applications.]

7.4.5.10.1 Backend computer

[An adjunct to a larger commercial computer (the host) which performs the actual database operations on request from the host.]

7.4.5.10.2 Database machine

[A specialized backend computer which supports a single data model partially or wholly in hardware.]

7.4.5.10.3 Associative machine

[A machine composed of logic elements attached to storage devices which can execute a specified operation (e.g. searching or comparing) in parallel.]

7.4.6 Management of database systems

[All aspects of control of the database to facilitate efficient access for approved use and to prevent unauthorized usage.]

7.4.6.1 Policy and legal aspects of security (see also 9.4.3 and 9.5.5.1)

[Assignment of responsibility for physical security and system correctness and the actions to take when there is a privacy violation.]

7.4.6.2 Database administrator (see also 9.6.1.6)

[The individual who is responsible for the management of the database for the benefit of the community of users.]

7.4.6.3 Centralization/decentralization (see also 9.5.6.1.1)

[The management issue which balances pressures for concentrating decision making and implementation at a single site and for distributing these responsibilities among numerous sites and users.]

7.4.6.4 Data auditing (see also 9.5.1.4)

[Techniques for periodic evaluation of the validity of the data in a database and the correctness of the data obtained as output from the system.]

7.4.6.5 Human factors

[The study of psychological factors important in the operation of database systems.]

PART IV. ANNOTATED TAXONOMY TREE
7. METHODOLOGIES
7.5 IMAGE PROCESSING

7.5 Image Processing (see also 7.9.3)

[The class of techniques for the manipulation and analysis of pictures by computers; also called picture processing.]

7.5.1 Digitization

[Conversion of pictures in their normal pictorial form into discrete-valued arrays of numbers, each number corresponding to a small element or pixel of the original picture.]

7.5.1.1 Sampling

[The brightness or the color is sampled at a discrete set of points to obtain analog signals corresponding to the gray levels (a measure of brightness) or the colors at those points.]

7.5.1.2 Quantization

[The analog signals are converted to discrete values representing the gray level or color.]

7.5.2 Compression (coding) (see also 3.3.2 and 3.3.3)

[A more compact representation of a digitized picture is obtained by coding, by eliminating redundancy or by approximation of the picture in order to achieve more efficient storage or transmission.]

7.5.2.1 Efficient encoding

[Special codes are used to represent pictures compactly while also permitting exact reconstruction.]

7.5.2.1.1 Shannon-Fano-Huffman (see 3.3.2.2.1)

7.5.2.1.2 Run length

[Representation of picture rows as sequences of runs having a constant value.]

7.5.2.1.3 Contour

[Representation of constant-value picture regions by specifying their boundary curves.]

7.5.2.2 Predictive coding

[Use of picture values on preceding rows or of preceding pixels on the current row to predict the next value on the current row, and application of compression techniques to the prediction errors.]

7.5.2.3 Transform coding

[Application of compression techniques to an invertible transform (Fourier, etc.) of a picture.]

7.5.2.4 Adaptive coding

[Switching from one compression technique to another

7. METHODOLOGIES

in accordance with statistical properties of parts of a picture.]

7.5.2.5 Interframe coding

[Compression of a sequence of pictures, taking advantage of redundancies between successive frames.]

7.5.3 Enhancement

[Methods for the improvement of picture quality.]

7.5.3.1 Grayscale manipulation

[Enhancement by increasing contrast (i.e. by spreading the gray levels further apart on the gray scale).]

7.5.3.2 Geometric correction

[The digital data is processed to compensate for geometrical distortions in the original picture.]

7.5.3.3 Smoothing

[Methods for reducing the effects of noise in the original picture.]

7.5.3.3.1 Noise cleaning

[Noise patterns are detected and deleted from the digitized picture.]

7.5.3.3.2 Averaging

[Smooths the data to decrease variability by averaging values at neighboring points or from independent pictures of the scene.]

7.5.3.4 Sharpening, deblurring

[In order to reduce the effects of blurring in the original picture.]

7.5.3.4.1 High-emphasis filtering

[Deblurring by emphasizing rapid fluctuations of gray level in a picture.]

7.5.3.4.2 Unsharp masking, Laplacian processing

[Deblurring by subtracting an intentionally blurred version of the picture from the picture itself.]

7.5.4 Restoration

[Compensation for known or estimated degradations affecting a picture.]

7.5.4.1 Inverse filtering

[Compensation for blur by frequency-domain filtering.]

7.5.4.2 Wiener filtering

[Frequency domain filtering to minimize the mean squared error between the original and restored pictures.]

7.5.4.3 Pseudoinverse restoration (see also 5.2.4.1.5)

[Least squares restoration using matrix pseudoinverse techniques.]

7.5.4.4 Recursive (Kalman) filtering

[Linear least squares techniques applied to a noisy picture.]

7.5.5 Reconstruction (from projections)

[The process of recovering (from projections) an image (picture) from (experimentally) available integrals of its grayness (value) over thin strips (or lines).]

7.5.5.1 Summation methods

[Methods for estimating the grayness of the image at any point by summing up the values of the integrals for all strips (or lines) which contain that point.]

7.5.5.2 Transform methods

[The Radon transform maps a function of two variables, say ℓ and θ, which represent a parameterized line, into another function Rf of two variables such that $[Rf](\ell,\theta)$ is the integral of f along the line parameterized by ℓ and θ; transform methods for reconstruction are numerical approximations of the inverse Radon transform.]

7.5.5.3 Series expansion methods

[Methods which assume that every image can be approximated as a linear combination of a fixed finite set of basis images; by taking many projections a large enough linear system is accumulated for the estimation of the coefficients in the linear combination.]

7.5.6 Segmentation

[Subdivision of a picture into regions for purposes of description.]

7.5.6.1 Thresholding

[Segmentation by grouping picture points on the basis of, for example, their gray levels.]

7.5.6.2 Edge detection

[Extraction of region boundaries using derivative or difference operators.]

7.5.6.3 Matching

[Detection of specified patterns or shapes by comparison with templates.]

- 7.5.6.4 Line detection

  [Detection of thin lines or curves by local operations.]

- 7.5.6.5 Tracking

  [Extraction of edges or curves by sequential search from one portion to the next.]

- 7.5.6.6 Region growing, partitioning

  [Extraction of uniform regions by repeated merging or splitting.]

- 7.5.6.7 Shrinking, thinning

  [Reduction of extracted objects to isolated points or thin curves.]

7.5.7 Feature measurement

[The computation of values of particular properties for given regions in a picture.]

- 7.5.7.1 Geometrical properties, shape

  [The properties measured are connectedness, area, perimeter, diameter, compactness, symmetry, elongatedness, straightness, convexity, etc.]

- 7.5.7.2 Textural properties

  [Spatial frequency or statistical characteristics of the gray level distribution.]

- 7.5.7.3 Moments

  [Weighted sums of the spatial arrangement of gray levels are calculated.]

- 7.5.7.4 Projections, cross-sections

  [Properties derived from intersections of the picture with families of lines or curves.]

- 7.5.7.5 Invariants

  [Properties invariant under given types of transformations of the picture.]

7.5.8 Scene analysis (see also 7.2.8.2)

[Pictures are described in terms of parts, properties, and relationships.]

- 7.5.8.1 From single images

  [Use of one picture to infer three-dimensional information about a scene.]

7.5.8.2 Use of auxiliary information

[Such as time sequences of pictures, stereopairs, range data, controlled lighting.]

7.5.9 Applications (see also 7.6.3)

[Classes of pictures and scenes that are often the subjects of analysis.]

7.5.9.1 Document processing

[Character recognition.]

7.5.9.2 Medicine and biology

[Cytology, radiology.]

7.5.9.3 High-energy physics

[Particle trace detectors.]

7.5.9.4 Industrial automation

[Fabrication, inspection, robotics.]

7.5.9.5 Remote sensing

[Earth sciences, meteorology, environmental studies, astronomy.]

7.5.9.6 Reconnaissance

[Target recognition.]

7.5.9.7 Forensic sciences

[Fingerprint or face recognition.]

7.5.9.8 Line drawing processing (see also 7.6.3.2)

[Cartography, drafting.]

PART IV. ANNOTATED TAXONOMY TREE
7. METHODOLOGIES
7.6 PATTERN RECOGNITION

7.6 Pattern Recognition

[Automatic and man-machine analysis, description, identification, classification, and extraction of patterns in data.]

7.6.1 Models

[Paradigms proposed for machine recognition.]

7.6.1.1 Classificatory models

[Processes which attempt to designate into which among several classes the input pattern falls.]

7.6.1.1.1 Deterministic models

[Classification procedures valid in the absence of statistical variability in the input pattern.]

7.6.1.1.2 Stochastic models

[Paradigms based on assuming inherent statistical variability for observations within pattern classes and between pattern classes.]

7.6.1.1.2.1 Statistical feature extraction

[Transformation of the original observed data into composite features in order to enhance the separation between classes or to reduce the separation between patterns in a class.]

7.6.1.1.2.2 Statistical classification and discrimination

[Use of statistical decision theory, statistical discriminant analysis, stochastic approximation-learning algorithms, etc. to achieve classification.]

7.6.1.1.2.3 State space models

[Use of state-space graphs and branch and bound heuristic search to model statistical decision trees for hierarchical classification.]

7.6.1.1.3 Fuzzy set model

[The assumption is made that patterns can simultaneously belong to more than one class with varying probabilities; assigns classes using the developing calculus of fuzzy sets.]

7.6.1.2 Structural models

[Approaches based on defining primitive patterns and identifying allowable structures in terms of relationships among primitives and substructures which combine primitives.]

7.6.1.2.1 Segmentation and primitive identification

[Patterns are subdivided into substructures in which primitive patterns are then delineated.]

7.6.1.2.2 Formal models for structural description and generation

[Grammars, AND/OR graphs, functional equations, finite state models, etc., which serve as mechanisms to generate and/or describe structures in terms of identified primitives.]

7.6.1.2.3 Integrated segmentation and structural description

[Provides for feedback and interaction between the segmentation and structural description processes.]

7.6.1.2.3.1 A priori knowledge representations

[Approaches to representing a priori structural and other knowledge in the database for use in the segmentation-structural description processing.]

7.6.1.2.3.2 Data driven, non-canonical parsers

[Model driven and data driven, top-down and bottom-up and non-left-right parsers which allow integrated segmentation-structural description.]

7.6.2 Design methodology

7.6.2.1 Pattern analysis

[Whatever is known about problem at hand is used to guide the gathering of data about the patterns and pattern classes; the data is then subjected to a variety of procedures for inferring deterministic and probabilistic structures which are present in the data.]

7.6.2.1.1 Histograms, scatter plots, other graphics

[One and two dimensional graphical displays of data variation.]

7.6.2.1.2 Cluster analysis algorithms

[Routines which are supposed to help discover natural groupings in the data.]

7.6.2.1.3 Mappings

[Linear and nonlinear transformations of the data based on various rationales designed to bring out a variety of relationships in the data.]

7.6.2.1.4 Exploratory data analysis

[Box plots, stem and leaf plots, median polishing, analysis of variance, regression analysis, outlier analysis, etc.]

7.6.2.1.5 Density estimation procedures

[Methods for the estimation of univariate and multivariate probability distributions.]

7.6.2.2 Feature evaluation and selection

[Evaluating and selecting features to find those which are effective in discriminating between pattern classes, in characterizing patterns within a class, or which permit the generation or reconstruction of the original patterns.]

7.6.2.2.1 Distance measures and error bounds

[For a given set of available observations, estimation of the Bayes error (i.e. the intrinsic overlap between different categories) and development of computationally feasible bounds on the Bayes error probability.]

7.6.2.2.2 Subset selection

[Search procedures are used to select a good subset of features for a particular pattern.]

7.6.2.3 Classifier design

[Methods for the design of the classification logic which provides the class designations or the design of the generative mechanism which generates class descriptions.]

7.6.2.3.1 Single stage

[Methods to design M-way classifiers for M classes which make a combined use of all the features in one stage to give a decision among classes.]

7.6.2.3.2 Hierarchical classifiers

[Methods to design decision tables,

decision trees and statistical decision trees with deterministic and statistical features and classifiers at each node.]

7.6.2.3.3 Grammatical inference

[Methods to infer grammars, etc. from samples in order to develop structural description generators.]

7.6.2.4 Pattern classification experiments

[Use of design and test samples to estimate classification performance of proposed features and classifiers.]

7.6.2.4.1 Using labelled samples

[Known samples from each class are used to design classifiers.]

7.6.2.4.2 Using unlabelled samples

[Using a labelled design sample and an unlabelled test sample to get performance estimates.]

7.6.3 Applications

7.6.3.1 Waveform analysis

7.6.3.1.1 Biomedical

[EKGs, EEGs, carotid artery pulse waves, etc.]

7.6.3.1.2 Speech

[Isolated word recognition, continuous speech recognition.]

7.6.3.1.3 NMR spectroscopy patterns

[Chemical compound identification.]

7.6.3.1.4 Other

[Waveforms from infra-red, seismic, radar, sonar, magnetic, multispectral and other sensors, time series, etc.]

7.6.3.2 Line drawings

7.6.3.2.1 Optical character recognition

[Document processing, page reading.]

7.6.3.2.2 Chemical structure diagrams

[Automatic input of chemical diagrams for computer processing.]

7.6.3.2.3 Blueprints, maps

[Engineering, drafting, cartography.]

- 7.6.3.2.4 Displayed mathematical expressions

    [Automated reading, input, and processing of mathematical formulas.]

- 7.6.3.2.5 Other

    [High energy particle traces, fingerprints, etc.]

7.6.3.3 Images (see 7.5.9)

7.6.3.4 Others

[Surveys and questionnaires; psychological, economic, marketing, etc.]

7.6.4 Implementations

[Elements of hardware and software used in designing and realizing pattern recognition systems.]

7.6.4.1 Hardware

- 7.6.4.1.1 Input

    [Electro-optical, laser and other passive and active sensors, tablets, recorders etc. to scan the physical environment and transform it into data.]

- 7.6.4.1.2 Processors

    [General and special digital and analog systems for computation, for preprocessing, feature extraction, selection, classification logic.]

- 7.6.4.1.3 Displays

    [CRTs, printers, hard copy devices etc. for one, two and three-dimensional displays of input, feature spaces, classification surface projections, decision trees, etc.]

7.6.4.2 Software

- 7.6.4.2.1 Packages for data analysis

    [Statistical packages, optimization routines, decision tree design and graphical displays.]

- 7.6.4.2.2 Design of interactive systems

    [Software systems for man-machine interactive design of pattern analysis and classification systems or for on-line recognition of patterns.]

PART IV. ANNOTATED TAXONOMY TREE
7. METHODOLOGIES
7.7 MODELING AND SIMULATION

7.7 Modeling and Simulation (see also 2.4.2.1.1.1 and 4.1.3.3.1)

[A model is an image of a system or object, used to gain experimental insight; simulation is the process of computerizing and using a model for its intended purpose.]

7.7.1 Characteristics of simulation models

7.7.1.1 Type

[The model may either represent a discrete event system or a continuous system.]

7.7.1.2 Dynamic

[The property that reflects that the system changes state as a function of the independent variable, "time."]

7.7.1.3 Mathematical/nonmathematical

[The model may be composed of mathematical objects such as numbers, algorithms, functions, equations, sets, but may instead or also be composed of nonmathematical objects, such as characters, names, etc.]

7.7.1.4 Deterministic/nondeterministic

[The model may incorporate certain variables, parameters, etc. which are either deterministic or are influenced by random processes.]

7.7.2 Discrete event simulation models

[The model exhibits the property of changing state at discrete increments of time and is considered by the observer to remain fixed between time increments.]

7.7.2.1 Purpose

[This type of model usually describes a system whose purpose is resource allocation and/or traffic operation in which queues may be formed and serviced and traffic congestion may occur; the object of the simulation is to enable experimental observation (and counting) of resource contention or queue performance, etc.]

7.7.2.2 System characteristics

7.7.2.2.1 Scope

[The model may represent an entire system or just a subsystem; the level of detail used in describing the system is often dependent on the scope.]

7.7.2.2.2 Transient behavior

[Concerned with the behavior of the system from initiation of operation until a (statistical) steady state is reached.]

7. METHODOLOGIES

7.7.2.2.3 Entities

[System entities, or objects, may be created at system initiation, or by the model itself; these may be permanent or may appear or disappear during the life of the system being simulated.]

7.7.2.2.4 Attributes

[Terms, parameters and symbols which identify or describe entities and which enable their grouping.]

7.7.2.2.5 Variable characteristics

[Properties pertaining to the nature and generation of variables.]

7.7.2.2.5.1 Random/deterministic

[Whether random processes affect variable values.]

7.7.2.2.5.2 Single-valued/functional

[Whether a variable may or may not take on a range of values.]

7.7.2.2.5.3 Exogenous/endogenous

[Whether a variable is generated from within or from outside the system.]

7.7.2.2.5.4 Independent/dependent

[Whether one variable is dependent on another; "time" is always an independent variable.]

7.7.2.2.5.5 Global/local

[Whether a variable affects the entire model or just one or more subsystems.]

7.7.2.2.6 Time advance

[How the "clock" in the model is updated.]

7.7.2.3 Simulation languages

7.7.2.3.1 World-view

[This concerns the aspect of the simulation process focused on by the language; it can greatly influence the design of the model.]

7.7.2.3.1.1 Transaction-oriented

[Primary entities are

7.7 MODELING AND SIMULATION

transactions (users, customers); these are processed through the system of resources and queues; GPSS (General Purpose Systems Simulator) is a well known example of this class.]

7.7.2.3.1.2 Event-oriented

[Primary entities are events when resources are allocated and used; records are kept of resource activity; Simscript is a well known example of this class.]

7.7.2.3.1.3 Process-oriented

[Primary focus is on the processes being simulated rather than on specific transactions or events; Simula is an example of this class.]

7.7.2.3.2 Language orientation

[Pertains to how coding of the model is accomplished.]

7.7.2.3.2.1 Statement-oriented

[Main entities of language are statements similar to those in more general purpose languages.]

7.7.2.3.2.2 Block-oriented

[Simulations are coded mainly through an alphanumeric shorthand which compresses instruction sets into shorthand "blocks."]

7.7.2.3.2.3 Other

[Hybrids of above plus special purpose programs used to effect a specific type of model.]

7.7.2.4 Applications

7.7.2.4.1 Business and industry

[Used for manufacturing facility design, resource allocation and scheduling, supply and distribution, etc.]

7.7.2.4.2 Government and social

[Used for military logistics and battle

7. METHODOLOGIES

simulation, ground and air traffic control, health services administration, etc.]

7.7.2.4.3 Engineering and science

[Used for computer hardware and software design, telecommunication system design, engineering project planning and control, system failure and repair analysis, etc.]

7.7.3 Continuous system simulation models

[The model exhibits the property of changing state as a continuous function of time.]

7.7.3.1 Purpose

[This type of model usually describes a system which undergoes dynamic change of state as a result of a physical or chemical perturbation or stress; the object of such a model is to enable the experimental observation (and measurement) of transient and steady state behavior of the system; the model usually obeys physical/chemical laws and can be expressed in terms of differential and algebraic equations, or in certain cases, operational transform notation.]

7.7.3.2 System characteristics

7.7.3.2.1 Scope

[As with discrete simulation, the model may represent an entire system or may be partitioned into subsystems.]

7.7.3.2.2 Transient behavior

[Concerned with startup behavior of the system and, particularly, with system stability before a steady state is reached.]

7.7.3.2.3 Steady state behavior

[Behavior of the system during ongoing operation; stability is again a particular concern.]

7.7.3.2.4 Variable characteristics

[Properties pertaining to nature and generation of variables.]

7.7.3.2.4.1 Random/deterministic (see 7.7.2.2.5.1)

7.7.3.2.4.2 Single-valued/functional (see 7.7.2.2.5.2)

7.7.3.2.4.3 External/internal

[The variable value may be internally determined or may

7.7 MODELING AND SIMULATION

be imposed by an interface with the external world such as a boundary condition.]

  7.7.3.2.4.4 Independent/dependent (see 7.7.2.2.5.4)

  7.7.3.2.4.5 Stable/unstable

[Whether the variable values remain bounded or become unbounded.]

7.7.3.2.5 Time advance

[How time is advanced in the model; usually dictated by the numerical integration technique used in the program.]

7.7.3.3 Simulation languages

 7.7.3.3.1 Type

[Pertains to the model description capabilities of the language.]

  7.7.3.3.1.1 General purpose

[Language or system has the capability of implementing most continuous systems; CSMP (Continuous System Modeling Program) is an example of this class.]

  7.7.3.3.1.2 Hybrid

[The capability of the language extends to mixed continuous/discrete systems; GASP IV (General Activity Simulation Program) is an example of this class.]

  7.7.3.3.1.3 Special purpose

[The language is designed only for specific continuous systems (e.g. DYNAMO for industrial dynamics).]

 7.7.3.3.2 Language orientation (see 7.7.2.3.2)

7.7.3.4 Applications

 7.7.3.4.1 Engineering and science

[Used for system design and testing and for safety, operation and failure analysis.]

 7.7.3.4.2 Business and industry

[Used for plant design, process control,

industrial dynamics, automation and economic analysis.]

7.7.3.4.3 Government and social

[Used for medical and physiological systems analysis; atmospheric, hydrodynamic, and pollution modeling; military weapon system analysis; command and control.]

7.7.4 Modeling considerations

[The factors important in the process of developing a model and also in the handling of inputs and outputs.]

7.7.4.1 Model validation and verification

[The actions taken by the modeller to ensure the model's validity with respect to reasonability and its veracity in relation to measured data.]

7.7.4.2 Random number generation

[Techniques for creating random variates used to introduce random effects in the system; see also 5.2.6.1.]

7.7.4.3 Model data reduction and analysis

[Techniques concerned with the statistical design and analysis of simulation experiments.]

7.7.4.4 Model portability and documentation

[How well the model can be moved from one computing environment to another and how well it is documented.]

PART IV. ANNOTATED TAXONOMY TREE
7. METHODOLOGIES
7.8 SORTING AND SEARCHING

7.8 Sorting and Searching

[Sorting is concerned with ordering records according to a field in the record called the key; searching is concerned with finding a record with a particular key in a (usually sorted) file.]

7.8.1 Sorting algorithms

[Algorithms for performing the sorting operation.]

7.8.1.1 Internal sorting

[The records to be ordered are contained entirely in main or virtual storage so that the sorting algorithm need not be concerned with record retrieval from external devices.]

7.8.1.1.1 Selection

[Successive phases of the ordering process find the record with the highest (or lowest) key of those still unsorted.]

7.8.1.1.1.1 Straight selection

[The largest key, then the second largest, etc., is found by sequential comparison and placed (via exchange) in the last, then second last, etc., position.]

7.8.1.1.1.2 Tree selection (Tournament)

[Keys are compared in pairs in successive stages until the largest (or smallest) number is found; successive runners-up are then identified, taking advantage of the elimination tree implicitly formed.]

7.8.1.1.1.3 Heapsort

[First a "heap," which is a complete binary tree in which the keys are ordered on any path from the root, is created; then this is sorted by successively moving the largest element to the root.]

7.8.1.1.1.4 Other

[Such as quadratic selection.]

7.8.1.1.2 Insertion

[Unsorted records are placed one-by-one into their proper position in the data structure holding already sorted records.]

7. METHODOLOGIES

7.8.1.1.2.1 Straight insertion

[The last several (up to all) records in an ordered sublist are physically moved one storage unit downward to accommodate a new record.]

7.8.1.1.2.2 Shellsort (Diminishing Increment Sort)

[Sublists defined by a diminishing interval between records are ordered until a final ordering with interval one involves the entire list.]

7.8.1.1.2.3 Linked linear list insertion

[Given an ordered linked linear list, the proper location for a new record is found by sequential scan from the head of the list and the new record is spliced in by appropriate change of link pointers.]

7.8.1.1.2.4 Tree insertion sort

[Already sorted records are organized into a binary tree of n nodes so that the proper place to insert any new key can be located in at most n log n probes.]

7.8.1.1.3 Exchange

[Unsorted records gradually approach their ultimate correct relative position through spatial exchange with other unsorted records.]

7.8.1.1.3.1 Partition exchange (Quicksort)

[Ordering by successively choosing ever smaller subintervals (partitions) within which keys are exchanged with respect to a pivot element until all keys to the left of the pivot are less than that pivot and all keys to its right are greater.]

7.8.1.1.3.2 Merge exchange (Batcher's parallel sort)

[Pairs of sorted subsequences are merged after obtaining those subsequences by exchanging pairs of non-adjacent records whose locations are chosen in accord with an intricate geometric pattern well adapted to parallel computation.]

7.8.1.1.3.3 Other

[Such as bubble sort and radix exchange.]

7.8.1.1.4 Enumeration (counting)

[Each key is compared with all others to determine how many are smaller and how many larger in order to determine where that key should be stored with respect to the top of the list.]

7.8.1.1.5 Merging

[Two or more ordered subfiles are combined into one ordered list in such a way as to take advantage of the initial partial ordering.]

7.8.1.1.6 Distribution

[Records are distributed into logical bins according to some characteristic such as constituent digits or numeric range; after each distribution the entire list is collected and the process repeated until a final collection produces an ordered list.]

7.8.1.1.6.1 Radix

[Records are distributed into partitions by comparing "digits" of their keys; when the radix is 10, the method is called "digital sorting."]

7.8.1.1.6.2 Address calculation

[The limiting case of radix sorting in which every potentially occurring key value has storage space reserved for it.]

7.8.1.2 External sorting

[Ordering of records in sets of a size that cannot be contained in a real or virtual address space; I/O operations on the set are explicit in the ordering algorithm.]

7.8.1.2.1 Direct access merge

[Ordering of elements using a merging process which places ordered subsequences of elements on disk, drum or a similar

7. METHODOLOGIES

device; subsequences are continually read, written and combined until a single sequence is formed.]

7.8.1.2.1.1 String management

[Algorithms which determine the placement of ordered strings of records on direct access devices.]

7.8.1.2.1.2 Space management

[Algorithms which allocate memory and storage spaces for buffers, work areas, procedures and the data itself.]

7.8.1.2.2 Tape merge

[Ordering of records using a merging process involving multiple tape units.]

7.8.1.2.2.1 Balanced

[In which the merge attempts to have equal numbers of ordered strings on each tape unit.]

7.8.1.2.2.2 Unbalanced

[In which the merging process distributes ordered strings by an algorithm which leads to unequal numbers of strings on tape units according to a predefined series such as the Fibonacci series.]

7.8.1.2.3 Distributive

[The use of tape or direct access devices to hold bins (areas) defined for distribution and collection of records placed by virtue of having a key within a range or having a particular key digit value.]

7.8.1.2.3.1 Record management

[Algorithms which address issues of record movement to and from defined areas.]

7.8.1.2.3.2 Space management

[Algorithms which address the definition of areas, the allocation of storage space to defined areas and the subdivision and redefinition of areas.]

7.8.2 Searching algorithms

[Algorithms for locating and retrieving records of a file.]

7.8.2.1 Sequential comparative

[Location by comparing a search argument against keys consecutively from the beginning of the list.]

7.8.2.1.1 Unordered list

[A list of unsequenced records which can be searched sequentially as efficiently as by any other method.]

7.8.2.1.2 Ordered list

[A list of sequenced records whose order enables various improvements over a sequential search.]

7.8.2.2 Tree-based comparative (see also 3.1.2.5)

[Location by definition of a tree structure such that each comparison eliminates a large number of records from further consideration.]

7.8.2.2.1 Binary search

[Use of a binary tree structure so that each comparison eliminates about one half of the records still to be inspected.]

7.8.2.2.2 Fibonacci search

[Use of the Fibonacci series to define the pattern of search.]

7.8.2.2.3 Higher order trees

[Use of m-way tree structures (where each parent has up to m offspring) for partitioning a list to be searched; common types are B-trees and trie structures.]

7.8.2.3 Classification argument

[Use of a specific part of a search argument to find an appropriate partition of a list; a common example is a dictionary search.]

7.8.2.3.1 Truncated index

[Use of a truncated form of the search argument to find an appropriate partition of a file; subsequent partitions may be formed by using subsequent digits going from left to right.]

7.8.2.3.2 Frequency ordering

7. METHODOLOGIES

[The search argument is compared sequentially to key values of a list ordered by the probability of occurrence of the keys.]

7.8.2.4 List structure

[For searching through records stored as lists linked to each other by pointers.]

7.8.3 Programming considerations

7.8.3.1 Data structures (see also 3.1)

[The organization of fields within records and also the encoding of the data in fields.]

7.8.3.2 File structures (see also 3.2.3)

[The logical and physical organization of files on storage media.]

7.8.3.3 Program structures

[The non-algorithmic overhead portion of sorting and searching programs.]

7.8.3.4 Hashing

[The logical address of the data to be accessed is computed by transforming the key by Boolean or arithmetic manipulation to an index, pointer or direct machine address.]

7.8.3.5 Buffering

[The use of main storage space to organize flow of records to and from auxiliary storage and input/output devices.]

7.8.4 Hardware organization

7.8.4.1 Processor

[Those attributes of a processing unit which influence the behavior of a sort or search program including data movement capability, comparison logic, comparison times, register structure, addressing structure.]

7.8.4.2 Storage

7.8.4.2.1 Storage mapping

[Mechanisms of a processor/memory combination which determine what data and procedures will be in main storage at any time and how they are addressed.]

7.8.4.2.2 Storage size

[The actual physical capacity of the main storage which determines how large a list may be subjected to internal sorts.]

7.8 SORTING AND SEARCHING

7.8.4.3 I/O subsystem

7.8.4.3.1 Device characteristics

[Attributes of auxiliary storage and I/O devices such as rotational delay, seek time, start/stop time, data transfer rate, physical block structures, address protocols, which determine the speed of external sorts.]

7.8.4.3.2 Overlap capabilities

[Attributes of an I/O subsystem which pertain to how much data flow can occur between main storage and peripheral devices during processing of other data.]

7.8.4.4 System architecture (see also 1.3)

7.8.4.4.1 Parallel processors

[Machines on which it is feasible to decompose a sorting process into a set of closely co-operating parallel processes.]

7.8.4.4.2 Multiprocessors

[Machines on which decomposition of a sort or search is feasible but not at the detailed level of a parallel processor.]

7.8.5 Algorithm analysis

[Tools and techniques used to predict the time and space efficiency of sorting and searching algorithms.]

7.8.5.1 Comparisons

[Number of comparisons of keys depends upon the algorithm and upon the distribution and permutation of the keys of the data submitted for sorting.]

7.8.5.2 Data movements

[Algorithms differ in the number of exchanges of record position or numbers of record transfers.]

7.8.5.3 Storage required

[Algorithms differ in the amount of main storage required to form ordered lists.]

7.8.5.4 Underlying theory

[Statistical and mathematical theory underlies the analysis of sorting; formal analysis involves a knowledge of information theory, probability theory and combinatorial mathematics.]

PART IV. ANNOTATED TAXONOMY TREE
7. METHODOLOGIES
7.9 COMPUTER GRAPHICS

7.9 Computer Graphics

[The computer generation and manipulation of two dimensional pictures (images) on hard copy or display devices, typically using operator controlled man-machine interaction.]

7.9.1 Hardware architecture

[The output devices on which the picture is plotted or displayed and the associated digital controller (display processing unit), with any associated input devices used for creating or manipulating pictures.]

7.9.1.1 Display devices (see also 1.4.2.3)

[Both hard copy devices for passive, output-only graphics and soft copy (display) devices for interactive graphics.]

7.9.1.1.1 Refresh devices

[Display devices used for interactive graphics based on cathode ray tube (CRT) technology; an image is "painted" (refreshed) some thirty times per second on the CRT surface from a stored digital representation (the "display file") of the picture.]

7.9.1.1.1.1 Vector displays

[The image is composed from a set of output primitives such as lines (vectors), conics, and characters drawn on a Cartesian grid - also called a calligraphic display; multiple levels of intensity and a few levels of color may be possible.]

7.9.1.1.1.2 Raster displays

[The images are specified as lines, characters and solid areas in grayscale (intensity) or color but are encoded as dots ("pixels") of varying intensity (grayscale or color) on a Cartesian grid ("raster").]

7.9.1.1.2 Image storage display devices

[Display devices which store the image itself electronically, as in a direct view storage tube's plasma panel, rather than storing a digital representation of the image.]

7.9.1.1.3 Large screen displays

[Large display surfaces used for group

7. METHODOLOGIES

viewing, typically with low resolution and low rates of interaction.]

7.9.1.1.4 Plotters

[Sheets of paper or plastic are held stationary (flatbed) or are moved (on tape or drum) while a drawing implement (pen or stylus) traces the drawing by being stepped along the two axes of the paper.]

7.9.1.1.5 Microfilm recorders

[A digitally (computer) controlled microfilm camera records the image displayed on a high precision CRT or an electron beam or laser may draw directly on the film.]

7.9.1.1.6 Screen copiers

[Devices which print a display screen's image on photosensitive paper.]

7.9.1.1.7 Special purpose hardware/firmware

[Hardware/firmware for providing additional output primitives or manipulations of output primitives.]

7.9.1.1.7.1 Curve generators

[Hardware for generating output primitives such as conics, or space filling two or three dimensional curves.]

7.9.1.1.7.2 Font/symbol generators

[Frequently used, application dependent special symbols (Greek letters, flowchart boxes, circuit components, etc.) are effectively added as output primitives to the "character set."]

7.9.1.1.7.3 Transformation hardware

[Allows geometric transformations such as translation, scaling and rotation of primitives.]

7.9.1.1.7.4 Clipping divider

[Produces the ("clipped") pieces of two or three-dimensional objects "in view" through a two-dimensional "window" or a three-dimensional "view volume" (rectangular pyramid).]

7.9.1.1.7.5 Hidden line/surface eliminator

[Equipment which computes which pieces of objects are obscured by others to produce a more realistic image.]

7.9.1.1.7.6 Other

[Such as area filling hardware or hardware which mixes pixels with live video.]

7.9.1.2 Input (interaction) devices

[Operator controlled devices whose values or changes of state are monitored by the display device's controller or computer to provide data for building or manipulating the picture or data structure or the image itself.]

7.9.1.2.1 Location-specifying devices

[Such as a data tablet with stylus, cursor crosshairs, joystick or trackball, typically used to provide x,y locations on the screen for on-line drawing.]

7.9.1.2.2 Picture element picking devices

[Such as a light pen, data tablet with comparator or "mouse" driven cursor used to identify a displayed element of the picture for further processing; alternatively a touch-sensitive display may be pointed to directly.]

7.9.1.2.3 Value-generating devices

[Such as control dials or keyboards used for specifying values to the applications program.]

7.9.2 Graphics systems (see also 2.3.1)

[Architecture and configuration.]

7.9.2.1 Single terminal systems

[Single display terminal, perhaps attached to a multi-access computer.]

7.9.2.2 Multi-terminal systems

[Multiple display terminals, perhaps shared by the same display controller, capable of simultaneous operation.]

7.9.2.3 Remote graphics

[Terminal located at a distance from the supporting host computer, connected via cable or telephone line.]

7. METHODOLOGIES

7.9.2.4 Satellite graphics

[Where supporting host, typically a minicomputer, is connected to a larger, time-shared mainframe for division of labor and resource sharing.]

7.9.2.5 Network graphics

[Multiple graphics system interconnected to multiple hosts via a computer communication network for more general division of labor or resource sharing.]

7.9.3 Picture (image) generation (see also 7.5)

[How objects themselves and the location of a "synthetic camera" in object space are described to the graphics system so that the system can generate a picture of the objects.]

7.9.3.1 Object description

[Specifying primitives and their attributes such as line style, color, etc.]

7.9.3.1.1 Linear segmentation of the display file

[The object space is divided into pieces, typically one per object, to allow pictures to be selectively manipulated (deleted, intensified, etc.) by manipulating the segment's stored representation in the refresh buffer.]

7.9.3.1.2 Segment hierarchy in the display file

[Segments may contain references to other segments in addition to primitives.]

7.9.3.2 Viewing operations

[Using varying combinations of either hardware or software to calculate the correct "view" of the object(s) described to the graphics system.]

7.9.3.2.1 Windowing

[Specifying the portion of the object space to be viewed.]

7.9.3.2.2 Clipping

[Extracting the part(s) in view, eliminating those not, to calculate the view.]

7.9.3.2.3 Projections

For three dimensions, projecting the viewed portions on a projection plane and then mapping them to the view surface, using projections such as perspective or axonometric.]

7.9.3.2.4 Viewports

[Describing where on the view surface the two-dimensional image is to appear.]

7.9.3.2.5 Boxing and extents

[Alternate methods for specifying mappings from window to viewport, especially useful in hierarchies.]

7.9.3.2.6 Image transformations

[Rotating, translating, or scaling a clipped, mapped (and projected) image of an object.]

7.9.4 Graphics utilities

[The programmer's interface with the graphics system.]

7.9.4.1 Graphics support software

[Input handling and picture making software.]

7.9.4.1.1 Graphical languages and extensions

[Languages or extensions to existing languages which allow picture creation and manipulation.]

7.9.4.1.2 Graphical subroutine packages

[Those added to existing languages, usually Fortran, to perform graphics functions.]

7.9.4.2 Picture description languages

[Formal languages for describing syntax of drawings.]

7.9.4.3 Typical application system design

[Integrated applications packages for building and manipulating objects and their graphical representations.]

7.9.4.3.1 Plotting packages

[For mapping data easily to graphical plots and charts possibly including automatic contouring, curve fitting, etc.]

7.9.4.3.2 Character and symbol packages

[Which build fancy character sets (fonts) or frequently used symbols.]

7.9.4.3.3 Graphics editors/drafting packages

[Packages which allow the user to build and modify pictures interactively.]

7.9.5 Object modeling

[Methods for representing and manipulating the relevant data

7. METHODOLOGIES

defining the two or three-dimensional objects.]

7.9.5.1 Object hierarchy

[Data representations using a hierarchy of objects, each composed of lower level objects and/or primitives.]

7.9.5.2 Object transformations

[The building blocks in an object hierarchy may be geometrically translated, rotated, scaled (and even clipped) prior to insertion in the higher level objects which use them.]

7.9.5.3 Model to picture transformation

[The application dependent process which describes a completed model to the graphics system for making a picture.]

7.9.6 Methodology and techniques

[Important concepts and techniques in graphics software.)

7.9.6.1 Device independence for portability

[Refers to the programmer's ability to specify output and input to and from high level logical graphic devices rather than to the actual physical devices on a given configuration, to achieve program transportability.]

7.9.6.2 Modeling versus viewing

[Separating the functions for the building and manipulation of objects (a function of the graphics system or the applications program) from those for taking pictures of them (always a function of the graphics system).]

7.9.6.3 Higher level versus basic ("kernel") graphics systems

[Analogous to high level language and assembly language in terms of high level features supplied versus fewer, possibly more device dependent, but more efficient features.]

7.9.6.4 Interaction techniques

[Commonly accepted mechanisms for obtaining operator commands and data to change object parameters on the picture.]

7.9.6.4.1 Menu picking

[Selecting choices from a list displayed by using a light pen or cursor.]

7.9.6.4.2 Zooming/panning

[Looking at a complex or large picture in (magnified) pieces, as if with a simulated video camera.]

7.9.6.4.3 Level of detail/extents

[Providing alternate, more or less detailed representations of picture pieces as a function of zoom level.]

7.9.6.4.4 Interrupt-driven dialogues

[The interactive program viewed as a finite state automaton, making state transitions and providing outputs/processing on the basis of operator input actions.]

7.9.6.4.5 Interactive recognizers

[A means of providing input in a dialogue via real time pattern recognition.]

7.9.6.5 Human factors engineering (ergonomics)

[The art and science of building powerful but easy and pleasant to use graphical man-machine dialogues.]

7.9.6.5.1 Workstation organization

[How user friendly the physical terminal is.]

7.9.6.5.2 Operator interface

[How easy to use, forgiving, reliable, powerful, responsive, etc. the applications package is.]

7.9.6.5.3 Programmer interface

[How easy to use, reliable, powerful, well-documented, etc. the hardware and support software are.]

7.9.7 Three-dimensional graphics and realism

[Special problems of representing complex real or imaginary solid objects.]

7.9.7.1 Hidden line/surface elimination

[Mathematical algorithms for determining which lines/surfaces are obscured and therefore cannot be displayed from a given viewpoint in three-dimensional object space.]

7.9.7.2 Surface representation

[Mathematical models such as polygon nets and parametric patches for smoothly varying curved surfaces such as automobile bodies.]

7.9.7.3 Texture, shading and lighting models

[Simulating the effects of light sources on objects

with varying surface gray scale, color, texture, reflexivity, etc.]

7.9.7.4 Aliasing

[Problems of low resolution of current raster display devices which produces a staircase effect for non-vertical or non-horizontal lines and edges.]

7.9.7.5 Dithering

[A means for simulating gray scale on a two tone raster display.]

7.9.7.6 Animation

[Making moving pictures on-line or off-line, a frame at a time.]

7.9.7.6.1 Simple motion dynamics

[Using transformation facilities to translate, rotate, or scale pieces of the picture.]

7.9.7.6.2 Key frame animation

[Computer interpolation of corresponding points between two key frames to produce the in-between frames.]

7.9.7.6.3 Frame at a time

[The picture is recomputed, probably in pieces, for each frame, to give the most general transformations.]

7.9.7.6.4 Simulators

[Real time animation of realistic objects such as airplanes under operator control.]

PART IV. ANNOTATED TAXONOMY TREE

8. APPLICATIONS/TECHNIQUES [ILLUSTRATIVE]

PART IV. ANNOTATED TAXONOMY TREE
8. APPLICATIONS/TECHNIQUES
8.1 BUSINESS DATA PROCESSING

8.1 Business Data Processing

[Systems that support the various functional units of a business or administrative organization, such as those listed below (which are not exhaustive).]

8.1.1 Financial systems

[Systems that support the financial organization; generally, these are transaction-based, i.e., large numbers of individual transactions (either input, e.g., cash receipts, or output, e.g., invoices) are processed on a periodic basis.]

8.1.1.1 Accounting

8.1.1.2 Cash management

8.1.1.3 Asset management

[Such as depreciation calculations for fixed assets.]

8.1.2 Personnel systems

8.1.2.1 Skills inventory

[Allows matching organizational needs with personnel qualifications.]

8.1.2.2 Payroll

8.1.2.3 Employment statistics

[Such as those required by various governmental agencies in support of Equal Opportunity and Affirmative Action programs.]

8.1.3 Marketing systems

8.1.3.1 Sales forecasting

8.1.3.2 Sales analysis

8.1.3.3 Order entry

[Whereby incoming orders from customers are processed and the necessary follow-up paperwork is produced.]

8.1.4 Distribution

[Systems that aid an organization in distributing its product or resources to its customers.]

8.1.4.1 Vehicle routing

[Systems that optimize the deployment of vehicles, e.g. trucks, in the distribution of merchandise between warehouses and retail outlets.]

8.1.4.2 Load balancing

[Systems that optimize the distribution to customers of a resource such as electrical power or natural gas.]

8. APPLICATIONS/TECHNIQUES (ILLUSTRATIVE)

8.1.5 Manufacturing systems

8.1.5.1 Requirements planning

[Systems that forecast materials and parts required to support production schedules.]

8.1.5.2 Production scheduling

8.1.5.3 Inventory control

[Systems that manage the physical status and cost of raw material, work-in-process, and finished-good inventories.]

8.1.5.4 Quality control

[Systems that monitor the quality of production so that departures from standards can be quickly identified and corrective action taken.]

8.1.6 Management information systems

[Systems that support management by processing information extracted from the transaction-oriented systems listed above).]

8.1.6.1 Budget analyses

8.1.6.2 Product profitability analyses

8.1.6.3 Cash flow analyses

8.1.7 Miscellaneous

8.1.7.1 Purchasing

[Systems that track the status of open purchase orders and the performance of vendors.]

8.1.7.2 Stockholder records

[Systems that pay dividends, maintain stockholder mailing lists, transfer shares, etc.]

PART IV. ANNOTATED TAXONOMY TREE

8. APPLICATIONS/TECHNIQUES

8.2 SCIENTIFIC AND ENGINEERING DATA PROCESSING TECHNIQUES

8.2 Scientific and Engineering Data Processing Techniques

8.2.1 Data acquisition

[Systems in which the computer is an on-line device for the collection of experimental data.]

8.2.1.1 Passive instrumentation

[Devices which record data production without participating in any other way with the process being monitored.]

8.2.1.1.1 Analog devices

[Such as, but not limited to, strip charts or acoustically recorded magnetic tape.]

8.2.1.1.2 Digital devices

[Such as, but not limited to, punched paper tape or cards or digitally recorded magnetic tape.]

8.2.1.2 Computer-controlled experiments (see also 2.3.4)

[Experimental situations controlled by a computer which feeds back intermediate data in such a way as to guide the controlled equipment in producing output data of more direct interest.]

8.2.1.2.1 Special-purpose controllers

[Processes are controlled by a special purpose computing device whose function cannot be changed after installation.]

8.2.1.2.2 General-purpose controllers

[Processes are controlled by a general purpose digital computer which can be reprogrammed to meet changing conditions.]

8.2.2 Analysis of experimental data

8.2.2.1 Data refinement and presentation

[The processing of raw data to improve its quality without substantial reduction in its quantity.]

8.2.2.1.1 Smoothing and filtering (see also 5.1.6.2.8 and 7.5.3.3-4)

[The elimination of extraneous noise and insignificant or undesirable fluctuations in value.]

8.2.2.1.2 Graphical presentation (see also 7.9)

[The display of raw or refined data as a graph or histogram either transiently on a CRT device or permanently on a medium suitable for direct human viewing.]

8. APPLICATIONS/TECHNIQUES (ILLUSTRATIVE)

8.2.2.1.3 Transformation of domain

[A systematic change of independent variable to a new domain such as, for example, transformation from a time domain to a frequency domain via Laplace or Fourier transformation.]

8.2.2.2 Error analysis (see also 5.3.1.3)

[The calculation and ascription to each raw data point of an estimated or maximum limit of error based on the assumption that it is drawn from a population of events conforming to a known frequency distribution such as the uniform, normal, Poisson or bimodal distributions.]

8.2.2.3 Statistical analysis (see also 5.2.6)

[The attempt to derive interrelationships among data points through application of certain classical statistical methods.]

8.2.2.3.1 Analysis of variance

[A method of analyzing differences among means of sets of samples.]

8.2.2.3.2 Multiple correlation

[A method for determining the degree to which each of several random variables can be expressed as a linear combination of all of the others.]

8.2.2.3.3 Stratified sampling

[A method for estimating the stability of a group mean through comparison with the means of subgroups (strata) into which the group is divided.]

8.2.2.3.4 Other statistical analyses

[Such as factor analysis, Bayesian estimation, cross-tabulation, and regression analysis.]

8.2.2.4 Data reduction

[The replacement of a set of experimental data, possibly many hundreds or thousands of individual readings, with a far smaller number of parameters from which close approximations to the original data can be recovered when the derived parameters are used in conjunction with the assumed functional dependence governing the phenomena being studied.]

8.2.2.4.1 Least squares curve fitting (see also 5.1.6.2.1 and 5.1.7.1)

[A method which, given an assumed functional dependence containing several

8.2 SCIENTIFIC AND ENGINEERING DATA PROCESSING TECHNIQUES

free parameters, yields values for these parameters which minimize the sum of the squares of the differences between each actual data point and its corresponding value calculated from the assumed functional dependence.]

8.2.2.4.2 Spline function analysis (see also 5.1.6.1.5 and 5.1.6.2.6)

[A curve fitting method where dissimilar functional relationships may be assumed on successive intervals of the independent variable's range and where contiguous segments of the fitted curve (splines) are joined smoothly at interval boundaries.]

8.2.2.4.3 Rational approximation (see also 5.1.6.2.5)

[Curve fitting based on a functional dependence consisting of a ratio of polynomials both of whose coefficients are varied, often heuristically, until the maximum error in the fit is less than some desired tolerance.]

8.2.2.4.4 Orthogonal function expansion

[A method whereby the data under study is assumed to be suitably described by a finite sum of orthogonal mathematical functional terms whose coefficients are derived from the data.]

8.2.2.4.5 Extraction of parameters

[Replacement of raw data by a small number of statistical parameters which characterize it such as low and high values, average, median, standard deviation.]

8.2.3 Simulation and model building (see also 7.7)

[A body of techniques whereby either a complex natural phenomena (such as weather) or a man-made artifact (such as an oil refinery or a nuclear reactor) is modeled by a limited set of equations or contingent events which, when followed computationally, track the essential features of the object under study throughout some reasonable period of observation, for the purpose of design, prediction or training.]

8.2.3.1 Design automation

[Where the purpose is to optimize the function or aesthetics of a manufactured object or engineered system or structure.]

8.2.3.2 Simulation for prediction

[Where the purpose is to predict the actual occurrence of events such as weather, earthquakes or economic development, etc.]

8. APPLICATIONS/TECHNIQUES (ILLUSTRATIVE)

8.2.3.3 Simulation for training

[Where the purpose is to provide realistic training to human operators prior to or supplementary to their being given control of the actual device or situation for which they will be responsible such as in an aircraft or ship simulator.]

8.2.4 Theoretical analysis

[The development and manipulation of mathematical formulas or algorithms which form the basis for calculating the theoretical behavior of a phenomenon or device.]

8.2.4.1 Arithmetic calculation (see also 5.)

[Analysis whose object is the production of a sequence of numbers which may ultimately be compared with experimental data.]

8.2.4.1.1 Solution of equations (see 5.1.1, 5.1.3 and 5.1.4)

8.2.4.1.2 Solution of inequalities (see 5.2.5)

8.2.4.1.3 Monte Carlo simulation (see 5.2.6.2)

8.2.4.2 Algebraic manipulation (see also 7.1)

[Where the computer manipulates directly the algebraic expressions which describe a physical process.]

8.2.4.2.1 Data representation (see 7.1.1)

8.2.4.2.2 Manipulative techniques (see 7.1.2)

8.2.4.2.3 Symbolic languages (see 7.1.3)

PART IV. ANNOTATED TAXONOMY TREE
8. APPLICATIONS/TECHNIQUES
8.3 COMPUTER-ASSISTED INSTRUCTION

8.3 Computer-Assisted Instruction

[Use of computers to present, often in the context of a dialog, drills, practice sessions and tutorial sequences to students.]

8.3.1 Hardware

8.3.1.1 Systems

[From single-user, personal computers to large time-sharing systems.]

8.3.1.2 Communications

[Between user terminal and the computer system.]

8.3.1.3 User interface

8.3.1.3.1 Typewriter keyboard

8.3.1.3.2 Graphic display (see also 7.9.1)

8.3.1.3.3 Pointers

[Including a light pen and, perhaps also, a joystick or touch panel.]

8.3.1.3.4 Audio input

[Such as words, phrases and digits which can be recognized by the computer (see also 1.4.1.7 and 7.2.4.1).]

8.3.1.3.5 Audio output

[From computer-controlled tape cassettes, speech synthesizers or other audio response devices.]

8.3.2 Software

8.3.2.1 Operations

[Maintains and processes records on student performance as well as delivering the instruction itself and handling special requirements.]

8.3.2.2 Development

[Of the instructional material itself; includes drafting and revising material, assembling graphics, processing text and checking logic.]

8.3.2.3 Utilities

[Such as for collection of data, analysis of data and scheduling of students.]

8.3.3 Applications

8. APPLICATIONS/TECHNIQUES (ILLUSTRATIVE)

8.3.3.1 Skills practice

[In, for example, basic arithmetic manipulations.]

8.3.3.2 Diagnostic testing

[Attempts to diagnose gaps in knowledge or understanding by presentation of appropriate test items or item sequences.]

8.3.3.3 Tutorial instruction

[By which new material is presented and its understanding tested.]

8.3.3.4 Simulation (see also 7.7)

[By which simulated situations are presented to the student (e.g. trajectories of rockets) in order to present material and test understanding of it.]

8.3.3.5 Modeling (see also 7.7)

[By which students build and test models based on data presented to them (e.g. fitting polynomials to data).]

8.3.3.6 Problem solving

[In mathematics, science or engineering.]

8.3.3.7 Artistic creation

[Of graphic arts and even music and written composition.]

8.3.3.8 Materials production

[Of graphical, animated and written materials where the computer and its display provides assistance to the author.]

8.3.4 Authoring

[Refers to the mechanisms by which authors produce CAI materials to be used by students.]

8.3.4.1 Question-and-answer sequences

[The most commonly used CAI technique.]

8.3.4.2 Simulations and models

[Involves the development of more complex and sophisticated modes of presentation of material; see 8.3.3.4 and 8.3.3.5.]

8.3.4.3 Information structures

[Which provide a basis for automatic generation of tutorial or test sequences or which may supply an information base for students to explore from their terminals.]

8.3.5 Evaluation

    8.3.5.1 Cost

    8.3.5.2 Effectiveness

    [Of learning achieved by student.]

    8.3.5.3 Reliability

    [Of hardware, software, communications and the results achieved.]

    8.3.5.4 Acceptance

    [By both teachers and students.]

8.3.6 Implications

    8.3.6.1 Learning

    [Including accomplishments which appear in other subjects, later in the same subject (without computer assistance) or in general learning.]

    8.3.6.2 Social interaction

    [Relative to more traditional modes of instruction.]

    8.3.6.3 Access to information

    [Other school information or more general information for the citizen and consumer.]

PART IV. ANNOTATED TAXONOMY TREE
8. APPLICATIONS/TECHNIQUES
8.4 TEXT (WORD) PROCESSING

8.4 Text (Word) Processing

[Creation and manipulation of character data, typically for preparation of documents such as manuals, contracts, proposals, books, newspapers, magazines and computer programs.]

8.4.1 Text editing

[Creation and alteration of the content of the document, usually interactively.]

8.4.1.1 Program editors

[For entering or correcting programs, typically by specifying commands for inserting, deleting, or replacing character strings on a program-line-by-line basis.]

8.4.1.1.1 Line editors

[Operate on one line of text at a time, usually referred to by a line number.]

8.4.1.1.2 Context editors

[Operate on lines or larger portions of the text; references are either by line number or by context patterns (e.g. all occurrences of one word are changed to another).]

8.4.1.2 Manuscript editors

[For non-line-oriented, more freeform manuscripts containing tables, equations or even pictures.]

8.4.1.2.1 Word processing systems

[Typically single terminal, stand-alone systems for producing simple documentation without photocomposition output.]

8.4.1.2.2 Shared logic systems (clusters)

[More sophisticated minicomputer-based systems with up to several dozen terminals and photocomposition output.]

8.4.1.2.3 Time-shared systems

[Shared-logic systems based on a large main frame with still more sophisticated features.]

8.4.1.3 Interaction/command specification

[The subsystem with which the user instructs the text editing system about the operations to be performed.]

8.4.1.3.1 Command language editors

[The user specifies commands as functions with parameters; usually used on hard copy

8. APPLICATIONS/TECHNIQUES (ILLUSTRATIVE)

and simple video terminals.]

- 8.4.1.3.2 Screen editors

   [Commands are specified primarily by function keys; parameters are specified by moving cursors or by devices such as joysticks.]

- 8.4.1.4 On-line manuscript exploration facilities

   [Where documents may be created especially to be browsed or read on-line, not just in hard copy manuscript form.]

8.4.2 Output creation

[Producing the text on a hard copy output device.]

- 8.4.2.1 Graphic arts/typography

   [Art and science of fostering communication through aesthetic, functional arrangements of text and graphics on the page.]

- 8.4.2.2 Mark-up systems

   [Composition of characters into lines which are justified, possibly hyphenated, and contain special effects such as font changes, tabbed text, equations, etc.; typically based on mark-up codes inserted in the text stream.]

   - 8.4.2.2.1 Hyphenation and justification

      [For automatic splitting of words at ends of lines and for producing flush right (justified) margins.]

   - 8.4.2.2.2 Tabular composition

      [Tools for convenient representation of tabular data.]

   - 8.4.2.2.3 Mathematical formulas

      [For displaying mathematical equations, including special symbols, subscripts and superscripts.]

   - 8.4.2.2.4 Format/typesetting codes

      [For "marking up" the text, that is, specifying paragraphing, indentation and other page formatting parameters as well as special symbols needed to process text for output on printers and typesetting machines such as photo and video composers.]

- 8.4.2.3 Layout and pagination system

   [Composition of lines into aesthetically pleasing pages, possibly containing multiple columns, tables, pictures, etc.]

- 8.4.2.3.1 Noninteractive/automated pagination

  [Off-line, algorithmic layout of simple, standard formats (e.g. for books).]

- 8.4.2.3.2 Interactive layout and pagination

  [Heuristic layout of more complex material using operator feedback (e.g. for newspaper pages).]

- 8.4.2.3.3 Page makeup terminals

  [Special interactive graphics systems capable of showing text and line drawings, possibly even halftones, for interactive layout.]

8.4.3 Auxiliary services

[Additional facilities provided by the computer to augment or replace manual processes: table of contents and index generation, spelling correction, journalling of changes and version maintenance, electronic mail for notification and distribution, teleconferencing and collaboration, workflow management, etc.]

- 8.4.3.1 Routing, queueing and electronic mail

  [Messages or documents are sent from one user or location to another within a system or via a message switching network between systems.]

- 8.4.3.2 Picture handling (graphics)

  [Inclusion of line drawings, halftones (gray scale) or color pictures within the text.]

- 8.4.3.3 Automatic index generation

  [Author specified keywords are searched for on each page as it is processed for subsequent sorting and printing.]

- 8.4.3.4 Spelling verification

  [Techniques for automatic detection and correction of spelling errors, typically using a dictionary.]

- 8.4.3.5 Journalling and version management

  [The system logs changes by author, date, time and other information to allow auditing, file backup, reconstruction or alternate versions.]

8.4.4 Input

[Subsystem for capturing raw text (typically with embedded typesetting codes) via key-to-disk, personal computers, on-line terminals or OCR.]

PART IV. ANNOTATED TAXONOMY TREE
9. COMPUTING MILIEUX

PART IV. ANNOTATED TAXONOMY TREE
9. COMPUTING MILIEUX
9.1 THE COMPUTER INDUSTRY

9. Computing Milieux

[Those aspects of computing that deal primarily with the interface between scientific and technological aspects of computing and the non-technical (or external) aspects.]

9.1 The Computer Industry

[Includes quantitative and qualitative factors relating to the computing industry, its markets and suppliers.]

9.1.1 Statistics

[Quantitative data about the computer industry.]

9.1.1.1 Employment

[Data on the number and types of individuals employed in the industry.]

9.1.1.2 Installation

[Data on computer installations, such as the quantity (by model number) of each unit installed; includes both hardware and software; also includes statistics on computer utilization.]

9.1.1.3 Financial (see also 9.1.3)

[Financial data, such as sales and net income of manufacturers.]

9.1.2 Markets

[Quantitative data about the markets for products and services sold by the computer industry.]

9.1.2.1 Product

9.1.2.1.1 Hardware (see 9.1.3.1)

9.1.2.1.2 Software (see 9.1.3.2)

9.1.2.1.3 Computing services (see also 9.1.3.3, 9.1.3.5 and 9.1.3.6)

[Such as contract programming or facilities management.]

9.1.2.1.4 Communications services (see also 9.1.3.4)

[Such as those used to transmit data between dispersed locations.]

9.1.2.2 Geographical

9.1.2.2.1 Domestic
9.1.2.2.2 Foreign

9.1.3 Suppliers

[Quantitative data and qualitative information on the computer industry suppliers of products and services.]

9.1.3.1 Hardware

    9.1.3.1.1 CPUs

    [Main frame manufacturers; often supply many of the other items under 9.1.3.1 and 9.1.3.2.]

    9.1.3.1.2 Peripheral equipment

    [Such as disks, drums, tapes and mass storage devices.]

    9.1.3.1.3 Terminals

    [Such as keyboard-printer terminals and visual display units (CRTs).]

    9.1.3.1.4 Data entry systems

    [Such as keypunch, key-to-disk, optical character recognition or voice recognition equipment.]

    9.1.3.1.5 Communications

    [Such as modems for connecting terminals to communications lines.]

    9.1.3.1.6 Other

    [Such as computer-output-microfilm (COM).]

9.1.3.2 Software

    9.1.3.2.1 System software

    [Such as operating systems and language processors.]

    9.1.3.2.2 Application software packages

    9.1.3.2.3 Custom software

    [For applications unique to one customer.]

    9.1.3.2.4 Other

    [Such as utilities, database management systems, etc.]

9.1.3.3 Computing services (see also 9.5.4.1)

    9.1.3.3.1 Service bureaus

    [Which supply a variety of computer services to customers, usually by batch processing (see 2.2.1) or remote job entry (see 2.2.2).]

    9.1.3.3.2 Time-sharing (see 2.2.3.1)

    9.1.3.3.3 Data banks

[Which give customers access to (usually specialized) databases; see also 7.3.6 and 7.4.]

        9.1.3.3.4 Other

[Such as those involving specialized hardware or software, e.g. graphics.]

9.1.3.4 Communications services

        9.1.3.4.1 Packet switching (see 3.4.2.2.2)

        9.1.3.4.2 Common carriers

[Usually using the normal telephone network.]

9.1.3.5 Professional and management services (see also 9.5.4.2)

        9.1.3.5.1 Consultation

        9.1.3.5.2 Education and training

        9.1.3.5.3 Newsletters and manuals

        9.1.3.5.4 Facilities management

[Management of computer facilities by an outside vendor.]

        9.1.3.5.5 Placement and recruiting

        9.1.3.5.6 Maintenance

[Including both hardware and software.]

9.1.3.6 Leasing companies

[Which obtain hardware from manufacturers and lease it to third parties.]

PART IV. ANNOTATED TAXONOMY TREE
9. COMPUTING MILIEUX
9.2 EDUCATION AND COMPUTING

9.2 Education and Computing

   9.2.1 Education about computing

      9.2.1.1 Graduate level

      [Masters and doctoral programs.]

         9.2.1.1.1 Computer science

         [Programs focused primarily on software, computer systems, mathematical or more theoretical aspects of computer science and engineering.]

         9.2.1.1.2 Information systems

         [Programs focused primarily on applications of computer science and engineering emphasizing database and file systems, particularly in business.]

         9.2.1.1.3 Computer engineering

         [Programs focused primarily on the various hardware and associated systems aspects of computer science and engineering.]

         9.2.1.1.4 Service courses

         [Courses for non-computer science and engineering students.]

      9.2.1.2 Undergraduate level

         9.2.1.2.1 Computer science (see 9.2.1.1.1)

         9.2.1.2.2 Information systems (see 9.2.1.1.2)

         9.2.1.2.3 Computer engineering (see 9.2.1.1.3)

         9.2.1.2.4 Computational mathematics

         [Programs, sometimes joint with mathematics departments, which focus on the interface of computer science and mathematics such as numerical analysis, discrete mathematical structures and probability.]

         9.2.1.2.5 Minor programs

         [Typically a coherent sequence of three to five computer science and engineering courses to be used as a portion of a student's program.]

         9.2.1.2.6 Service courses (see 9.2.1.1.4)

         9.2.1.2.7 Computer literacy (see also 8.7)

         [Courses concentrating on the history of computing, computer applications and the philosophical and social consequences of computing.]

- 9.2.1.3 Junior and community colleges

  - 9.2.1.3.1 Programmer trainee programs

    [Programs focused primarily at preparing students for entry level programming jobs.]

  - 9.2.1.3.2 Data processing programs

    [Programs focused primarily at preparing students for data entry and computer operations jobs.]

  - 9.2.1.3.3 Service courses (see 9.2.1.1.4)

  - 9.2.1.3.4 Computer literacy (see 9.2.1.2.7)

- 9.2.1.4 Secondary schools

  - 9.2.1.4.1 Programming instruction

    [Instruction in elementary aspects of computer science and engineering primarily focused on programming.]

  - 9.2.1.4.2 Computer literacy (see 9.2.1.2.7)

  - 9.2.1.4.3 Data processing instruction

    [Career education primarily focused at the preparation of computer operators or data entry operators.]

- 9.2.1.5 Continuing education

  - 9.2.1.5.1 Seminars and short courses

    [Programs to allow the professional to gain additional competence in a special area in a limited period of time.]

  - 9.2.1.5.2 Self-assessment

    [A procedure by which a professional takes a test in privacy and is referred to key readings or other materials to overcome deficiencies identified by the test.]

  - 9.2.1.5.3 Certification (see also 9.6.4.1)

    [A procedure whereby an individual takes a test which, when successfully passed, will result in the issuing of a certificate of competence.]

9.2.2 Educational uses of computers (see also 8.3)

- 9.2.2.1 Higher education

  [Includes computer-assisted instruction, computer-managed instruction, computer simulations, and the use of computers in college and university administration.]

9.2.2.2 Precollege education

[Includes uses of computer-assisted instruction for such things as drill and practice in grammar schools and general instruction, as well as introduction of basic concepts and use in school administration (see also 9.2.2.1).]

9.2.2.3 Continuing education

[For remedial instruction and aids to the handicapped (see also 9.2.2.1).]

9.3 HISTORY OF COMPUTING

PART IV. ANNOTATED TAXONOMY TREE
9. COMPUTING MILIEUX

9.3 History of Computing

    9.3.1 Origins (pre-twentieth century)

        9.3.1.1 Beginnings

            9.3.1.1.1 Abacus

                [Which dates from the 5th century B.C. or earlier and is still widely used in the Orient.]

            9.3.1.1.2 Slide rules

                [First developed and used about 1650; widely used until recent years.]

            9.3.1.1.3 Napier's bones

                [A system of rods invented in the early 1600s by John Napier, the inventor of logarithms, which embodied a scheme for multiplication derived from an Arabic method.]

        9.3.1.2 Early calculators

            9.3.1.2.1 Pascal's calculator

                [Adding machine; dates from 1642; intended to be practical but quite unreliable.]

            9.3.1.2.2 Leibniz' calculator

                [Developed in 1670s to do all four arithmetic operations; notable for its stepped wheel mechanism.]

            9.3.1.2.3 Other early calculators

                [In 17th to early 19th centuries associated with names such as Schickard, Vaucanson and de Colmar.]

        9.3.1.3 Charles Babbage

            [1791-1871; the "father" of modern computers whose ideas were way ahead of the technology needed to implement them.]

            9.3.1.3.1 The difference engine

                [A device intended by Babbage to mechanize the production of mathematical tables using differencing techniques.]

            9.3.1.3.2 The analytical engine

                [Intended as a mechanical, automatically-sequenced, general-purpose calculator; embodied many ideas of modern digital computers.]

9.3.1.4 Calculators in the nineteenth century

9.3.1.4.1 Burroughs

[Which developed the first successful adding and listing machine in 1892.]

9.3.1.4.2 Odhner

[The Ohdner wheel mechanism was used in the Brunsviga and other European calculators.]

9.3.1.4.3 Other calculators

[Such as the Comptometer invented by Felt in 1884.]

9.3.2 Early developments (until 1950)

9.3.2.1 Punch card machines

[Which used analogs of IBM cards as the mechanism for storing and transferring data.]

9.3.2.1.1 Early developments

9.3.2.1.1.1 Hollerith

[As employee of U.S. Census Bureau, inventor of punched-card data processing and first machines for it; later founded company which became IBM.]

9.3.2.1.1.2 Powers

[Engineer at Census Bureau who developed system competitive with Hollerith's; later founded company which became Remington Rand.]

9.3.2.1.1.3 Use in census from 1890-1910

[First extensive use of punched-card data processing using Hollerith's system.]

9.3.2.1.2 Later punch card machines

[Sorters, tabulators, collators, reproducers; most sophisticated was the IBM Card Programmed Calculator of the early 1950s.]

9.3.2.2 Electromechanical and relay computers

9.3.2.2.1 Differential analyzers and other analog computers

[First one built by Vannevar Bush at MIT in 1931 followed by various other models; main purpose was to solve differential equations.]

- 9.3.2.2.2 Bell Telephone Laboratories relay computers

  [Six relay machines built in 1937-1949 by a team headed by George Stibitz and Ernest Andrews.]

- 9.3.2.2.3 Zuse Z3

  [Program-controlled calculator built in Germany in 1941; did not survive World War II.]

- 9.3.2.2.4 Mark I (IBM and Harvard U.)

  [Also called IBM Automatic Sequence Controlled Calculator; first large-scale, automatic computer; developed by Howard Aiken at Harvard; first placed in operation in 1944.]

- 9.3.2.2.5 Mark II - Harvard relay computer

  [A relay machine built by Aiken and completed in 1946.]

- 9.3.2.2.6 IBM SSEC

  [First computer built by IBM as successor to Mark I; installed in 1948.]

9.3.2.3 Early electronic computers

- 9.3.2.3.1 Vacuum tube flip-flops (Eccles-Jordan)

  [A high speed electronic switching device whose invention made the development of electronic computers possible.]

- 9.3.2.3.2 Atanasoff's computer project

  [First electronic calculating device developed at Iowa State University to solve simultaneous linear equations; never actually completed.]

- 9.3.2.3.3 Colossus

  [Developed during World War II in Britain for cryptographic purposes.]

- 9.3.2.3.4 The ENIAC

  [First electronic automatic computer developed by John Mauchly and J. Presper Eckert at Moore School of Electrical Engineering at University of Pennsylvania; first in operation in 1946.]

- 9.3.2.3.5 The EDVAC

[Successor to ENIAC; first stored-program electronic computer although preceded in operation by Cambridge University EDSAC in 1949; first operational in 1951; designed by John Von Neumann and others.]

- 9.3.2.3.6 Stored program computers

   [Such as the EDSAC and Whirlwind at MIT.]

9.3.3 Recent history

- 9.3.3.1 The first generation (vacuum tube computers)

  - 9.3.3.1.1 Early large-scale computers

     [Such as the UNIVAC 1 and the IBM 700 series.]

  - 9.3.3.1.2 Medium-scale computers

     [Such as the IBM 650, the Burroughs Datatron and the UNIVAC 80 and 90.]

  - 9.3.3.1.3 Magnetic core memory computers

     [Such as the RCA BIZMAC, the IBM 704 and 705 and the UNIVAC 1100 series.]

  - 9.3.3.1.4 First generation software systems

    - 9.3.3.1.4.1 Assemblers

       [Notably SOAP on the IBM 650 and SAP (later FAP) on the IBM 704 and 709.]

    - 9.3.3.1.4.2 Compilers

       [First algebraic language compiler on MIT Whirlwind; first business compiler and other software innovations on Univac I; these led to compilers on various other machines.]

    - 9.3.3.1.4.3 Other

       [Such as the Fortran Monitor System on the IBM 704 and 709.]

- 9.3.3.2 The second generation

   [Characterized by the replacement of vacuum tubes by transistors and by the use of magnetic core memories.]

  - 9.3.3.2.1 Early transistorized computers

     [Such as Lincoln Laboratory TZ-0, IBM 608, NCR304, RCA501 and Philco 1000.]

  - 9.3.3.2.2 Major second generation computers

[Such as the Philco TRANSAC, IBM 7090, UNIVAC 1107 and Control Data 1604 and 3000 series.]

9.3.3.2.3 Early virtual memory systems

[Notably the Atlas system in England.]

9.3.3.3 The third generation

[Characterized by the use of integrated circuitry in most (but not all) hardware, by the use of large operating system software and multiprogramming, and by the use of disk units as the medium for the storage of most data files.]

9.3.3.3.1 Hardware (see also 1.)

9.3.3.3.1.1 Integrated circuits

[In which a number of electronic components are on a single fabricated chip of semiconductor material.]

9.3.3.3.1.2 IBM 360 and 370 series

[The most widely used computers of the late 1960s and 1970s.]

9.3.3.3.1.3 Other major computer systems

[Such as the UNIVAC 1108 and 1110, Burroughs 6700 and 7700, Honeywell 6000 series, Digital Equipment PDP-10, Control Data 6000 and Cyber series and the Amdahl 470 V/6.]

9.3.3.3.1.4 Large-scale integration

[In which the number of components in an integrated circuit may be as high as 10,000 and the "board" itself is a silicon chip about 1/10" on a side.]

9.3.3.3.1.5 Minicomputers

[Such as the DEC PDP series and the Data General Nova series; see also 1.1.1.2.]

9.3.3.3.1.6 Microprocessors

[Many models made by such manufacturers as Intel, Motorola, Zilog, Texas Instruments and Hewlett-Packard; see also 1.1.1.1.]

9. COMPUTING MILIEUX

9.3.3.3.2 Software (see also 4.)

9.3.3.3.2.1 Operating systems

[Such as IBM's OS 360, UNIVAC's EXEC 8 and Control Data's SCOPE; see also 4.2.1.]

9.3.3.3.2.2 Time-sharing systems

[From CTSS and Multics developed at MIT and the Dartmouth Basic system to many commercial systems now available; see also 2.2.3.1.]

9.3.3.3.2.3 Other software systems

[Including many language processors, database management systems and various utilities; see also 4.2.2.]

9.3.3.3.3 Systems

9.3.3.3.3.1 Communications systems

[Networks such as the ARPA net and satellite systems for connecting remote computers; see also 2.2.4.]

9.3.3.3.3.2 Array processors

[Such as the ILLIAC IV.]

9.3.3.3.3.3 Vector processors

[Such as the Control Data Star and Cray 1 compters.]

PART IV. ANNOTATED TAXONOMY TREE
9. COMPUTING MILIEUX
9.4 LEGAL ASPECTS OF COMPUTING

9.4 Legal Aspects of Computing

[Facets of CS&E which may require interaction with or help from lawyers and courts.]

9.4.1 Software protection mechanisms

9.4.1.1 Copyright

[Federal statutory protection for the written or printed expression of a program.]

9.4.1.2 Patent

[Federal statutory protection for "apparatus and processes"; applicability of existing patent law to programs currently not clear.]

9.4.1.3 Trade secret

[State protection which forbids the disclosure of programs by unauthorized persons.]

9.4.1.4 Registration

[A copy of the program would be deposited with a Registrar to be kept secret but which would permit disclosure of the concept of the program.]

9.4.1.5 Contract

[Between parties who agree to keep a program confidential.]

9.4.2 Privacy

[Confidentiality of data about individuals in computer-based systems.]

9.4.2.1 Data banks

[Personal data in machine readable form in database systems; see also 7.4.]

9.4.2.2 Transmission systems

[Where personal data in machine readable form is transmitted between computers or between computers and terminals; see also 2.2.4.]

9.4.2.3 Transborder data flow

[Regulation of transfers of personal information across national boundaries.]

9.4.3 Security (see also 9.5.5.1)

[The protection of data from unauthorized access in computer-based systems.]

9.4.3.1 Civil sanctions

[Resulting from invasion of privacy, breach of

contract and unfair competition.]

9.4.3.2 Criminal sanctions

[Resulting from copyright violation, violation of computer crime laws, program, hardware or data theft, vandalism, violations of laws on privacy.]

9.4.4 Taxation

9.4.4.1 Hardware

[Such as investment tax credit and sales, use or property tax.]

9.4.4.2 Software

[See 9.4.4.1; status of most types is currently unclear.]

9.4.4.3 Data processing services (see 9.4.4.2)

9.4.5 Contracts

9.4.5.1 Hardware acquisition

9.4.5.1.1 Sales contracts

[All rights are transferred to the purchaser.]

9.4.5.1.2 Lease contracts

[The lessee obtains the right to use the equipment for a specified period of time sometimes with an option to purchase at the end of the period.]

9.4.5.2 Software acquisition

9.4.5.2.1 Sales contracts

9.4.5.2.2 Lease contracts

9.4.5.3 Maintenance

[May be with the vendor or a third party.]

9.4.5.4 Services

9.4.5.4.1 Processing

[Typically with a service bureau.]

9.4.5.4.2 Facilities management

[A firm's computer facilities are managed by another firm.]

9.4.5.4.3 Licenses

[Most common form of software acquisition in which the licensee obtains the right to use the software for a period and typically agrees not to make unauthorized copies or disclosures.]

9.4.5.4.4 Escrow of software in source form

[Where a third party has access to the source program so that, in case of emergency or failure of a licensor to remain in business, a licensee has an alternative to the original licensor.]

9.4.5.5 Other

[Such as for various kinds of insurance.]

9.4.6 Computer-communications interface

[All systems consisting of both computer and communications facilities; see also 3.4.]

9.4.6.1 FCC inquiries

[Into the interconnection of data communications equipment with common carrier facilities.]

9.4.6.2 Datapaths

[Use of other than common carrier facilities for data communications.]

9.4.6.2.1 Satellites

[As operated by INTELSAT (in which the U.S. representative is COMSAT) for a planet-wide commercial satellite system.]

9.4.6.2.2 Specialized common carriers

[Such as microwave, wire, cable television and laser communications.]

9.4.7 Antitrust

9.4.7.1 Government litigation

[For violations of the Clayton and Sherman Acts.]

9.4.7.2 Private suits

[By competitors for alleged violation of the Clayton and Sherman Acts.]

9.4.8 Other legislation/regulation

9.4.8.1 Import/export

[Concerned with such matters as custom duties and export licenses.]

9.4.8.2 Procurement

[Particularly regulations of the General Services

Administration for government purchases of products and services.]

9.4.8.3 Banking/EFT (see also 2.3.4.6.3)

[Regulations pursuant to the Financial Institutions Regulatory Act of 1978.]

9.4.8.3.1 Automated teller machines

[A particular form of a CBCT (see 9.4.8.3.2) which is a branch of a bank.]

9.4.8.3.2 Customer bank communication terminals

[Such as cash machines; regulated by both state and federal law.]

9.4.8.4 Personnel

9.4.8.4.1 Labor relations

[Such as unionization of data processing personnel.]

9.4.8.4.2 Licensing

[A form of proposed regulation or standardization of levels of competence of data processing personnel.]

9.4.8.5 Consumer credit

[Computerized systems such as those affected by the Fair Credit Reporting Act, the Fair Credit Billing Act and the Equal Credit Opportunity Act.]

9.4.9 The computer in litigation

9.4.9.1 Admissibility of computerized records

[Primarily concerned with exceptions to the hearsay rule of evidence for computerized records.]

9.4.9.2 Discovery of computerized records

[Refers to matters related to the discovery of evidence in computerized records under the Federal Rule of Evidence.]

PART IV. ANNOTATED TAXONOMY TREE
9. COMPUTING MILIEUX
9.5 MANAGEMENT OF COMPUTING

9.5 Management of Computing

9.5.1 Management support

[Functions which support the technical services and operations of the computing organization.]

9.5.1.1 Long range planning

[Of computer resources, including hardware, systems software, applications, personnel, facilities and financial.]

9.5.1.1.1 Capacity

9.5.1.1.2 Applications

9.5.1.1.3 Personnel

9.5.1.1.4 Financial

9.5.1.1.5 Facilities

9.5.1.1.6 Communications network

9.5.1.2 Financial management

[Such as capital and operating budgets.]

9.5.1.2.1 Budgets

9.5.1.2.2 Cost analyses

9.5.1.2.3 Charges for services

[Refers to how users are charged for services provided.]

9.5.1.3 Installation standards

[Both their development and enforcement.]

9.5.1.3.1 Hardware

9.5.1.3.2 Software

9.5.1.3.3 Interfaces

[Between computers and between computers and other (possibly remote) devices.]

9.5.1.3.4 Documentation (see also 4.1.3.4, 9.5.2.2.5 and 9.5.5.4.4)

[The form and requirements for the documentation of all programs and systems.]

9.5.1.3.5 Communication

[Such as the protocols by which data will be transmitted between devices.]

9.5.1.4 Auditing (see also 9.7.5.4)

9. COMPUTING MILIEUX

[Includes financial audits and audits of security and compliance with privacy laws.]

      9.5.1.4.1   Requirements and methodologies

      9.5.1.4.2   External auditors

9.5.2   Application system development

[Those tasks, from project initiation through application system maintenance, and including project management, which constitute the "life cycle" of a computer system as it moves from concept to operational status.]

   9.5.2.1   Project initiation

[Steps taken at the outset of a project to assure that it meets a need and is likely to be successful.]

      9.5.2.1.1   Requirements definition

[By the user of what the system is required to do.]

      9.5.2.1.2   Functional specifications

[Detailed specification of the system, including throughput, input and output.]

      9.5.2.1.3   Feasibility studies

[Which attempt to determine if the state-of-the-art, the laws of economics, and the organizational setting will permit successful completion of the project.]

   9.5.2.2   System development

      9.5.2.2.1   Design

[Detailed design of the system, including hardware configuration, program design and operational procedures.]

      9.5.2.2.2   Programming

      9.5.2.2.3   Module testing

[In which portions of the system, usually substantial ones, are tested for correctness and efficiency.]

      9.5.2.2.4   System testing

      9.5.2.2.5   Documentation (see also 4.1.3.4, 9.5.1.3.4 and 9.5.5.4.4)

      9.5.2.2.6   Acceptance testing

[By the user (or a third party) to see if a system meets its stated requirements.]

   9.5.2.3   Implementation

9.5.2.3.1 Conversion

[From whatever old system was in use to a newly developed one.]

9.5.2.3.2 Training

[Of personnel to use a new system.]

9.5.2.3.3 Post-implementation audit

[Which attempts to determine if the system is living up to expectations and, if not, why not.]

9.5.2.4 Application system maintenance (<u>see also</u> 9.5.5.4.3)

9.5.2.4.1 Error detection, reporting and correction

9.5.2.4.2 System enhancement

[In which new features are added or existing features are made more efficient or useful.]

9.5.2.4.3 Change control

[Which tries to ensure that changes to one part of a large system do not degrade performance or render inoperable other parts of the system.]

9.5.2.5 Project management

[Steps taken to ensure the timely completion of a project and to control costs and manpower.]

9.5.2.5.1 Schedule

9.5.2.5.2 Manpower

9.5.2.5.3 Cost

9.5.3 Advanced development

[Activities largely of a technical nature and generally carried out by staff groups to assess the feasibility of technological developments that might be applicable across all (or many) of the applications in a given installation.]

9.5.3.1 Hardware

9.5.3.1.1 Evaluation

9.5.3.1.2 Role of minicomputers and microprocessors

9.5.3.1.3 Intelligent terminals

9.5.3.1.4 Interfaces

[Between computers, between computers and peripherals and between computers or peripherals and communications facilities.]

9.5.3.2 Software

    9.5.3.2.1 Operating systems

    9.5.3.2.2 Programming methodology (see 4.1.3)

    9.5.3.2.3 Communications network control

        [Development of software for controlling message routing, message accounting, network status monitoring, etc.]

9.5.3.3 User systems

    9.5.3.3.1 Database management (see also 7.4)

    9.5.3.3.2 Word processing (see also 8.4)

    9.5.3.3.3 Office automation

9.5.3.4 Delivery of services

    9.5.3.4.1 Networks (see 1.3.2.2.4)

    9.5.3.4.2 Distributed processing (see 2.1.2.1)

    9.5.3.4.3 Interactive, real time and batch (see 2.2.1 and 2.2.3)

9.5.4 User services

    9.5.4.1 Computing services (see also 9.1.3.3)

        9.5.4.1.1 Local

            [Services to users located at the central computer installation.]

        9.5.4.1.2 Remote

            [Services to users at other locations; see also 2.2.2.]

    9.5.4.2 Professional and management services (see also 9.1.3.5)

        9.5.4.2.1 Consultation

        9.5.4.2.2 Design and programming

            [Professional services performed under contract to design and program applications (and other) systems.]

        9.5.4.2.3 Education and training

9.5.5 Operations

    9.5.5.1 Security (see also 7.4.5.9 and 9.4.3)

        9.5.5.1.1 Physical

            [From fire, theft, vandalism, etc.]

9.5.5.1.2 Data

[From unauthorized manipulation or access.]

9.5.5.1.3 Program

[From unauthorized manipulation or access.]

9.5.5.1.4 System

[From any activity which could cause the system (hardware, software and data) to malfunction or otherwise operate improperly.]

9.5.5.2 Computer operations

9.5.5.2.1 Console

[From which the operator gives commands to the system and monitors its operation.]

9.5.5.2.2 Peripheral

[Such as mounting disks or tapes or supplying printers with paper.]

9.5.5.2.3 Remote

[Servicing terminals and other remote devices.]

9.5.5.3 Support operations

9.5.5.3.1 Data entry

[Capturing source data, usually on cards, tape or disk.]

9.5.5.3.2 Scheduling

[Of when each job will be executed.]

9.5.5.3.3 Dispatching

[In which jobs - programs plus data - are sent to computer operations for execution.]

9.5.5.3.4 Library

[Of application and other support programs available to the user.]

9.5.5.3.5 Quality control

[Various techniques used to verify that input data and output reports meet system specifications.]

9.5.5.4 Maintenance

9.5.5.4.1 Hardware

9.5.5.4.2 System software

[Introductory enhancements and modifications; correcting bugs.]

9.5.5.4.3 Application software (see also 9.5.2.4)

[Introductory enhancements and modifications; correcting bugs.]

9.5.5.4.4 Documentation (see also 4.1.3.4, 9.5.1.3.4 and 9.5.2.2.5)

[Keeping it current and correcting errors as they are found.]

9.5.5.5 Capacity management (see also 2.4)

[Short range planning and management of production capacity including main frames, peripherals and communications networks; see also 9.5.1.1.1.]

9.5.5.5.1 Performance measurement

[Including throughput rates, memory utilization, etc.]

9.5.5.5.2 Configuration management

[Planning and management of equipment configurations.]

9.5.5.5.3 Reliability/availability

[Matters relating to the reliability and availability of hardware and software and their effects on useful capacity.]

9.5.6 Organization (see also 9.7.3)

9.5.6.1 Philosophy

9.5.6.1.1 Centralization/decentralization issues

[Such as those concerned with how much of the computer operations of an organization should be centrally controlled (perhaps with remote processing allowed) and how much should be controlled by other (user) organizations.]

9.5.6.1.2 Cost center/profit center issues

[Such as the degree to which corporate computing centers must operate at a "profit."]

9.5.6.2 External relationships

[Those between the computing facility, its users and its superiors.]

9.5.6.2.1 Reporting relationships

[Between the computing organization and higher authority.]

    9.5.6.2.2 Use of advisory committees

    [To endorse (or assign) priorities and allocate resources.]

    9.5.6.2.3 User liaison

  9.5.6.3 Internal organization

    9.5.6.3.1 Structure

    9.5.6.3.2 Job descriptions (<u>see also</u> 9.6.1)

  9.5.6.4 Personnel management

    9.5.6.4.1 Personnel evaluation

    9.5.6.4.2 Professional development

    9.5.6.4.3 Career path planning

    9.5.6.4.4 Salary administration

PART IV. ANNOTATED TAXONOMY TREE
9. COMPUTING MILIEUX
9.6 THE COMPUTING PROFESSION

9.6 The Computing Profession

[Refers to all aspects of computing which identify it as a profession and distinguish it from other professions.]

9.6.1 Occupational titles

9.6.1.1 Computer scientist

[A person involved in computer research or advanced development.]

9.6.1.2 Computer engineer

[A person involved in design or development of computer hardware.]

9.6.1.3 Systems analyst

[A person who analyzes systems to be computerized and designs the system to be implemented on a computer.]

9.6.1.4 Programmer

[A person who implements a system on a computer or, more generally, anyone who writes a computer program.]

9.6.1.5 Computer operator

[A person involved with any aspect of the operation of a computer, usually of a large computer system.]

9.6.1.6 Other

[Such as keypunch operators, data entry clerks, database administrators (see also 7.4.6.2).]

9.6.2 Organizations

9.6.2.1 Technical societies

9.6.2.1.1 AFIPS

[The American Federation of Information Processing Societies; a federation of 13 societies all or some of whose members are directly involved with computing.]

9.6.2.1.2 ACM

[The Association for Computing Machinery whose members are mainly computer scientists and programmers.]

9.6.2.1.3 DPMA

[The Data Processing Management Association whose members are mainly systems analysts, business programmers and managers of these.]

9.6.2.1.4 IEEE Computer Society

[The Institute of Electrical and Electronic

Engineers Computer Society whose members are mainly computer engineers.]

9.6.2.1.5 BCS

[The British Computer Society, by far the largest technical society outside the United States, whose members span all areas of computing interests in the UK.]

9.6.2.1.6 IFIP

[The International Federation of Information Processing whose membership consists of the computing society or other designee from about 40 countries.]

9.6.2.1.7 Other

[Such as the American Society for Information Science, the Canadian Information Processing Society and the Association Française pour la Cybernetique, Economique et Technique.]

9.6.2.2 User groups

[Which focus on the users of the computers of a particular manufacturer.]

9.6.2.2.1 SHARE

[Users of medium to large IBM computers; generally oriented toward scientific applications.]

9.6.2.2.2 GUIDE

[Users of medium to large IBM computers; generally oriented toward business data processing.]

9.6.2.2.3 USE

[Users of large UNIVAC computers.]

9.6.2.2.4 VIM

[Users of large Control Data computers.]

9.6.2.2.5 DECUS

[Users of Digital Equipment Corporation computers.]

9.6.2.2.6 Other

[Such as COMMON and CUBE.]

9.6.2.3 Trade associations

9.6.2.3.1 ADAPSO

[The Association of Data Processing Service Organizations; see also 9.1.3.3.1.]

- 9.6.2.3.2 CBEMA

    [The Computer and Business Equipment Manufacturers Association whose members are mainly manufacturers of computer hardware.]

- 9.6.2.3.3 Other

    [Such as the Computer and Communications Industry Association.]

9.6.2.4 ICCP

[The Institute for Certification of Computer Professionals which administers testing programs (see 9.6.4.1.1 and 9.6.4.1.2) to certify competence in various areas of computer science and engineering.]

9.6.3 Professional ethics

9.6.3.1 Canons

[The principles on which professional ethics are based.]

9.6.3.2 Considerations

[That will be taken into account in judging whether or not canons have been violated.]

9.6.3.3 Rules

[That will be followed in adjudicating alleged ethical violations.]

9.6.4 Testing programs

9.6.4.1 Certification (see also 9.6.2.4)

- 9.6.4.1.1 CCP

    [Certificate in Computer Programming which tests basic knowledge in programming with specialization in business programming, scientific programming or systems programming.]

- 9.6.4.1.2 CDP

    [Certificate in Data Processing which tests basic knowledge required for business data processing management.]

9.6.4.2 Self-assessment (see 9.2.1.5.2)

9.6.5 Licensing

[Is concerned with whether or not certain classes of computer professionals (e.g. programmers) should be licensed as, for example, accountants are.]

PART IV. ANNOTATED TAXONOMY TREE
9. COMPUTING MILIEUX
9.7 SOCIAL ISSUES AND IMPACTS OF COMPUTING

9.7 Social Issues and Impacts of Computing

9.7.1 Theory and methods

9.7.1.1 Theoretical perspectives

[Technological and social determinism, voluntarist approaches (e.g. symbolic interaction), pluralism, conflict theory.]

9.7.1.2 Historical analyses

9.7.1.3 Empirical methods

[Survey methods, experimental designs, unobtrusive measures, participant observation.]

9.7.1.4 Futures and forecasting

[Methods include Delphi, scenarios.]

9.7.1.5 Models for evaluating social effects

[Such as issues of what time scale to utilize, what parties to include in analysis.]

9.7.1.5.1 Utilitarian criteria

[Such as cost-benefit analyses.]

9.7.1.5.2 Non-utilitarian criteria

[Considerations of equity, deontological ethics -- ethical systems where the rules of conduct do not take the consequences of one's actions into account.]

9.7.2 Growth and development of computing

9.7.2.1 Diffusion of computing technology

9.7.2.2 Transfer of computing technology

[Between organizations.]

9.7.2.3 The computing world

[Studies of organizations and groups that develop and provide computer-based technologies and services - their social organization, ideologies, and interactions with clients/customers.]

9.7.2.3.1 Social organization

9.7.2.3.2 Ideologies

9.7.2.3.3 Expert-lay interaction

9.7.3 Organizations (see also 9.5.6)

9.7.3.1 Workplace

[Focuses on the work of people who use computer-based

technologies as an instrument in other work.]

- 9.7.3.1.1 Work styles and job characteristics
- 9.7.3.1.2 Employment and skills

    [Alterations in skill levels, career patterns.]

- 9.7.3.1.3 Productivity

    [Of individuals and work groups.]

- 9.7.3.1.4 Work organization

    [Social organization of the workplace; kinds of specialization, differentiation of work groups, integrating roles, scale of work organization.]

9.7.3.2 Organizational structure

- 9.7.3.2.1 Organizational power

    [How distribution of power is influenced by and influences the uses of computer-based technologies.]

- 9.7.3.2.2 Patterns of control

    [How computer-based technologies fit with patterns of (administrative) control within organizations.]

9.7.3.3 Decision making

- 9.7.3.3.1 Policy making

    [Problem finding, computing as a legitimizing instrument in bureaucratic politics.]

- 9.7.3.3.2 Decision style

    [Individual or group.]

9.7.3.4 Production of goods and services

9.7.3.5 Dependency upon computer-based services

9.7.3.6 Reliability of computer-based systems (see also 9.5.1.4)

[Including the auditability of data and models.]

9.7.4 Public policy issues and societal impacts

9.7.4.1 Public attitudes and perceptions of computing

9.7.4.2 Employment and labor markets

[Structural characteristics, new kinds of jobs, occupational mobility.]

9.7.4.3 Privacy and surveillance

9.7.4.4 Computing and electoral/legislative activity

    9.7.4.4.1 Voting

    9.7.4.4.2 Political campaign practices

        [Custom-tailored mailing lists and analyses of voters' preferences.]

9.7.4.5 Machine intelligence and human identity

    [Studies how concepts of self are influenced by beliefs about the capabilities of computing.]

9.7.5 Computing in industrial societies

9.7.5.1 Social organization

    9.7.5.1.1 Everyday life

    9.7.5.1.2 Inter-institutional and sectoral arrangements

        [Data sharing between different classes of institutions such as banks and credit-authorization firms.]

9.7.5.2 Publics and computer-using organizations

9.7.5.3 Service provision

    [How different publics influence the kinds of services provided with computer-based support, and the nature and impacts of the services provided.]

9.7.5.4 Accountability of computing

    [The extent to which different publics can hold providers of computer-based products and services responsible for what they produce; different models for accountability.]

9.7.5.5 Industrial growth

9.7.5.6 Computing and cultural values

    [Shifting emphasis, such as to quantitative forms of evidence.]

9.7.6 Cross-national impacts

9.7.6.1 Transborder data flow (see 9.4.2.3)

9.7.6.2 Computing in less developed countries

9.7.6.3 Multinational organizations and services

9.7.6.4 Technology transfer

    [Emphasizing transfer across national, political, and cultural boundaries.]

9.7.6.5 National sovereignty and balance of trade

[Ways that importing or developing computer-based products and services locally influences political and economic relations between different countries.]

9.7.7 Humanistic studies of computing

9.7.7.1 Philosophical inquiries

9.7.7.1.1 Ethical issues of computer use

9.7.7.1.2 Philosophical anthropology

[People's self-conceptions in light of computer-based technologies.]

9.7.7.1.3 Philosophy of technology

[Critical theory.]

9.7.7.2 Computing in literature

[Images of computing.]

9.7.7.3 Computing in the arts

[New modalities of composition such as computer music, animation, esthetics.]

V. CORE TERMINOLOGY

This section contains a number of common terms which are used without definition in the taxonomy tree or its annotations.

Address

[A name which labels a memory location or machine register.]

Algorithm

[A precisely defined sequence of steps for operating upon precisely defined input to solve a class of similar problems, producing a precisely defined output in a finite time while employing finite computational resources.]

Alphanumeric

[Pertaining to a character that might be either a letter, a number, or some special symbol or punctuation mark.]

Argument

[One of possibly several variables whose value fully or partially determines the value of the particular function with which it is associated.]

Auxiliary Storage

[Backup storage, such as that represented by drum, disk, or magnetic tape units, whose access time is longer than that of the computer's principal high speed memory but whose capacity is usually greater.]

Binary

[Pertaining to the base 2 number system.]

Bit

[A "binary digit"; a unit of information, either 0 or 1, in the base 2 number system.]

Buffer

[(v.) To process data to or from its ultimate destination by way of an intermediate segment of storage; in most modern machines, it is done to permit concurrency of computation and I/O processing; (n.) the memory segment used in a buffering operation; this may be as small as one bit or byte or word but is often several tens or hundreds of words in length.]

Bug

[An error in a computer program which either causes it to terminate prematurely or not at all or to calculate inaccurate answers.]

Byte

[A contiguous group of n bits, usually but not necessarily 8, where, n is typically a factor of the word length (e.g. four 8-bit bytes compromise one IBM 370 32-bit word).]

Cell

[The smallest unit of an addressable memory device that has a uniquely assigned retrieval address; can be as small as one bit (Burroughs 1700) or one byte (IBM 370) but is often the word length (CDC Cyber).]

Central Processing Unit (CPU)

[The operational heart of a digital computer, consisting of the control unit and arithmetic unit.]

Channel

[An information pathway connecting I/O units or auxiliary storage units to a digital computer; channels dedicated to only one (usually high speed) I/O device are sometimes called *selector channels*; those which serve several (usually low speed) devices concurrently are called *multiplexer channels*.]

Character

[One of the letters, digits, or special symbols that may be input to, output from, or stored in a particular digital computer.]

Code

[(v) To translate schematic logic, possibly in the form of a flow chart, into assembler- or procedure-oriented language statements; (n) a segment of a program or, in some industries, a synonym for program.]

Console

[That portion of a computing system which displays information concerning machine status and which allows communication between the computer and its human operator or maintenance engineer.]

Cycle Time

[The constant interval between successive beats of a master pulsed clock which is the minimum time during which any logical task can be performed by a computer; the execution times of each of its machine language instructions must then all be an integral multiple (perhaps 1) of the cycle time.]

Data Path

[The amount of information (usually expressed in bits) fetched from memory during one memory cycle; this is usually equal to a word length on medium-scale computers, a multiple of the word length on large-scale computers, and a factor of the word length on small computers.]

Debug

[To eliminate "bugs" or errors in a computer program through systematic checkout with test data.]

Delimiter

[A single character (often a space, comma, colon, period, semicolon, quote, or parenthesis) or a character string (such as **begin** or **end** in Algol) which denotes the beginning or the end ("delimits") of a particular programming language construct.]

Emulation

[The simulation of one computer on another through software whose execution efficiency is augmented by being able to use special purpose hardware features (firmware) specifically designed and built for such simulations.]

Expression

[A formula consisting of a sequence of one or more characters which represents a syntactically valid combination of identifiers and constants in a particular high level language and which can legally serve as the right-hand side of a replacement (assignment) statement.]

Field

[A portion of a record, having fixed or variable length, and assigned to hold a particular piece of information, e.g. an eleven-column card field assigned to hold a hyphenated social security number, or the first n bits of a particular machine language instruction which is assigned to serve as a command field.]

Flag

[A sentinel set during execution of a program indicating that a certain condition or error is present.]

Flow Chart

[A pictorial diagram, consisting of interconnected boxes and symbols of various shapes, which depicts the flow of control from start to finish of a particular process or computer program; "flow chart" is not synonymous with "algorithm," but drawing a flow chart is one way to document an algorithm.]

Hexadecimal

[Pertaining to the base 16 number system.]

Hollerith Card

[A basic source document for some present day information processing, consisting of a rectangular card capable of holding a 12 x 80 binary array of punch (1) or no-punch (0) combinations; the colums are numbered 1 to 80 from left to right; the rows are numbered 12, 11, 0, 1 . . . 9 from top to bottom; named for its originator, Herman Hollerith, a U.S. Census Bureau employee just prior to and after 1900.]

Instruction

[The elemental unit of a digital computer machine language program, consisting of a command and all necessary operand addresses.]

Iteration

[A repetitive calculation or manipulation usually involving the same operations each time on different data; a part of all significant computer programs.]

Job

[A single, self-contained task to be executed by a computer.]

Loop

[A sequence of instructions which is retraversed a finite number of times in order to apply the same transformation to successive elements of a data array or set of arrays.]

Machine Language

[That language consisting of instructions directly understandable to the computer without translation.]

Microsecond

[One-millionth of a second; 10^{-6} sec.]

Millisecond

[One-thousandth of a second; 10^{-3} sec.]

Nanosecond

[One-billionth of a second; 10^{-9} sec.]

Object Program

[The machine language program which results when a high level or assembly language program is translated by, respectively, a compiler or assembler; see also Source Program.]

Octal

[Pertaining to the base 8 number system.]

Operand

[A part of a machine language instruction which designates a quantity to be operated upon; also refers to a quantity in an arithmetic or logical expression which is not an operator.]

Operation Code (Op Code)

[In a machine language instruction designates the operation to be performed.]

Operator
[In a high level language designates the operation to be performed in an arithmetic expression (e.g. +, *) or logical expression (e.g. and, not).]

Procedure
[An independent program segment to perform a particular task, usually part of a larger program, often one needed many times by the program of which it is a part; *see also* Subroutine.]

Program
[A collection of machine instructions or source language statements which implement the logical sequence needed to solve a particular problem.]

Radix
[The base of a number system (e.g. the decimal system has radix 10).]

Register
[Any special purpose memory cell in a computer, usually distinct from main memory; examples include arithmetic registers, index registers, and special control registers.]

Response Time
[During interactive computing, the time between a human-activated request and the beginning of the computer's output in answer to that request.]

Roundoff Error
[The error which accumulates in the course of a calculation due to roundoff of intermediate results to the size of numbers allowed by the computer.]

Source Program
[A program written in a high level or assembly language which is translated (by a compiler or assembler) into machine language for actual execution on the computer; *see also* Object Program.]

Statement
[An individual high level language or assembly construct which expresses a complete thought; an analogous construct to a sentence in a natural language such as English.]

Subroutine
[A program segment to perform a particular, usually quite limited, task (e.g. compute a square root) which can be called (i.e. entered) from a main program (closed subroutine) or can be inserted where needed in the main program (open subroutine); *see also* Procedure.]

Throughput
[The total amount of useful work produced by a computer in a given time period.]

Turnaround Time
[The time between submission of a batch run and the availability of tangible output from that run.]

Variable
[An identifier (i.e. name) corresponding to a memory cell or cells whose contents are allowed to change during the course of a computation.]

Word
[A fixed length addressable unit of memory, usually the size required to hold a single-precision integer or floating point number.]

VI. UMBRELLA TERMS

This section contains definitions of terms used commonly in CS&E but which have such broad, sometimes amorphous, connotations that they are not natural nodes in our *Taxonomy* tree.

Applications Programming

[Programming activity whose objective is construction of programs which are of direct productive interest (such as payroll, billing, solution of scientific equations) as contrasted with systems programming activity whose objective is creation or maintenance of software such as operating systems and language processors which serve as tools for the applications programmer.]

Automation

[The production of a product or control of a process wholly or mostly by means of machinery in a situation that had formerly been human labor intensive. Since the term antedates modern digital or analog computation and control, the machinery referred to is not necessarily electronic or even electrical, though there would be few modern applications that do not rely on such technology to augment or control mechanical or hydraulic equipment that may also be involved.]

Cybernetics

[A term which once was a candidate to describe all of what is now called computer science but which is now used in a more restricted sense to connote automatic control theory and certain phases of artificial intelligence, principally those related to robotics. The term is derived from a Greek word meaning "steersman" or "helmsman" and derived its original impetus from the mathematical work of Norbert Wiener. Experimental work in the field makes heavy use of analog and hybrid computers as applied to continuous event simulation and continuous mathematics as contrasted with the emphasis on discrete event simulation and discrete mathematics characteristic of work with digital computers.]

Data Processing

[A generic term so broad as to be applicable to the entire range of mechanically-aided computational activity from those based on the slide rule or hand-held calculator up to those involving the fastest computers available. Without further qualification such as, e.g., *scientific* data processing, the term most often connotes business or administrative data processing wherein a large volume of input and output transactions are paramount and can be made with only a modest need for intermediate computational speed.]

General Purpose Computer

[A term once but no longer in common use which denoted a computer intended for all or almost all purposes for which a computer can be used in contrast to a special purpose computer (*see* node 2.3); whether or not a given computer system is general purpose or special purpose today is often a function only of how it is programmed and used rather than anything intrinsic in the hardware.]

Information Science

[Information science is an interdisciplinary science that investigates the properties and behavior of information, the forces that govern the flow and use of information, and the techniques, both manual and mechanical, of processing information for optimal storage, retrieval, and dissemination.]

Nonnumeric Programming

[Programming whose predominant component involves the manipulation or transformation of nonnumeric data structures, such as character strings, as contrasted with programming whose predominant component consists of conventional arithmetic operations applied to numeric data. Examples are language translation, compilation, textual information retrieval, computer-assisted instruction, and concordance generation.]

Software Engineering

[The attempt to design, produce, and maintain software with the same rigorous attention to verification of compliance with specification as typically attends the production of a turbine or other complex but carefully engineered structure which is expected to perform correctly when operated properly. Characteristic of the discipline applied to such efforts is the reduction of large programs into manageable modules; the heavy use of Structured Programming; and the use of "egoless" programming whereby members of the team examine one another's work in an effort to produce an error-free product.]

Standards

[In the U.S., data processing standards for both hardware and software are developed and published on a voluntary nongovernmental basis and coordinated through ANSI, the American National Standards Institute, a non-profit membership organization which is the U.S. member of the worldwide International Standards Organization (ISO). To date, standards exist for such high level languages as Fortran, Cobol, PL/I, APT, and for a wide variety of hardware-associated concepts such as magnetic ink character recognition (MICR), ASCII code, optical character recognition (OCR) fonts, and punched card and magnetic tape formats.]

Structured Programming

[A term with many meanings for which standard terminology has not yet been agreed upon. At its simplest, structured programming refers to the use of only three control structures in a program: (1) sequencing, (2) choice by using IF - THEN - ELSE, and (3) a looping construct which is usually represented by DO - WHILE. In some formulations, a CASE statement, or additional looping construct such as DO - UNTIL, is provided. In most uses of the term, the traditional unconditional transfer of control statement "GO TO" is either prohibited or strongly discouraged. Sometimes use of the term includes the concept of top-down (outside-in) design whereby overall program logic is divided into a small number of supervisory routines which call on a larger number of subprograms each of which is expected to consist of about a page of high level language statement logic or is itself further subdivided. Also sometimes included is the concept of stepwise refinement whereby any one subprogram is written in a series of progressively more detailed phases, the first of which is documented carefully but in paragraphs containing a high admixture of English, and the last of which consists entirely of high level language statements. This term is sometimes used as a catchall for such concepts as program verification and the chief-programmer team.]

Symbol Manipulation

[A term sometimes used to describe *everything* done by a digital computer since all computer operations may be interpreted as the manipulation of symbols stored in the memory of the computer. In a more restricted sense this term is used to describe (1) Nonnumeric Programming involving operations on alphanumeric character strings for any purpose, or still more narrowly, (2) programmed symbolic algebra wherein character strings representing polynomials or other mathematical structures are transformed to an alternative form according to formal rules such as those of integration, differentiation, or algebraic simplification.]

Systems Analysis

[A component of administrative planning activity occurring prior to program design and coding wherein all phases of data collection and transmittal are carefully designed from a human factors standpoint in an attempt to produce a smoothly functioning application.]

Systems Programming

[*See* definition and discussion under Applications Programming.]

VII. THE DEVELOPMENT OF THE TAXONOMY: PHILOSOPHY AND TECHNICAL ISSUES

Computer Science and Engineering, as defined in the Introduction, is a relatively new and rapidly changing field. The task of formulating a taxonomy for CS&E is therefore particularly subject to problems of achieving conceptual unity and stability. Conceptual unity for the overall taxonomic scheme is difficult to attain because there is no general agreement about the nature and structure of the discipline. Conceptual stability, i.e. validity of the taxonomic scheme over time, similarly suffers from the dynamic nature of the discipline itself.

The definition of CS&E which appears in the Introduction can, at most, be considered suggestive of the domain addressed. Ideally, appeal to the definition would allow an answer to the question—what is the range of topics included in CS&E? Once that question is decided a host of others arise, having to do with order, partitioning into subtopics, and the like, the objective being to derive a tree structure which includes all topics among its nodes, each at the appropriate level and in the appropriate relation to other topics. There are probably, in this context, advantages to a committee product over an individual one. The compromise inherent in the committee process tends to prevent much of the idiosyncrasy that an individual effort must almost surely exhibit even if it appears more conceptually unified.

With regard to conceptual stability, the test of time for a taxonomy such as this is not whether it will require change and update in the future—that is inevitable—but rather whether the present structure lends itself to consistent and gradual change as the discipline of CS&E develops. How the *Taxonomy* will stand up to this test is, of course, a very difficult question to assess. It will depend upon the pattern of change in computing theory and practice.

One reasonable measure, however, depends on how the probability that change will be required is related to the level of a node. In setting up the tree, it has been the intention to plan the higher level nodes so that they are not only important but fundamental, i.e. less subject to change as the field continues to develop.

Although discussion of the dynamics of a discipline is essentially an exercise in futurism, it can be hazarded that certain areas are more likely than others to have major changes among their nodes:

1. the effects of very large-scale integration (VLSI) technology have only just begun to impact the discipline, and a real potential exists for our view of Hardware, Systems, Data and Software (nodes 1-4) being radically altered in the not too distant future by this technology;

2. the Theory of Computation (node 6) is most of its aspects a young discipline (e.g. computational complexity would not even have appeared as a subtopic ten years ago) and unquestionably as more theoretical underpinnings, particularly for software-related areas, are developed, this node will undergo steady expansion.

Attempting to study any subject by studying its status at a single time is dangerous; thus political science without history is suspect. Because CS&E is so dynamic, all involved with this project see it as fraught with dangers. But it should perhaps be viewed at least as source material to the future historian of CS&E. Errors, misapprehensions, and lack of conceptual unity may be corrected—or, at least, mitigated—in the next version of this *Taxonomy*. The Committee has a strong conviction that a task such as this must be undertaken once in order that it can ever be done right.

The four nodes for Hardware, Computer Systems, Data, and Software may be taken as comprising the "core" of CS&E. Difficulties in constructing this part of the tree were of three kinds:

1. arriving at the most logical ordering of the four core topics;

2. the need to choose under which of these nodes to place particular topics;

3. the difficulty of finding a logical place anywhere under these four headings for some very important topics.

Although various different suggestions were made as to the most appropriate ordering of the nodes, a clear consensus among the Committee favored the choice given, using the rationale that Hardware and Software are the "pure" endpoints of a core computer science and technology continuum and that, while both the Computer Systems and Data nodes encompass hardware and software aspects, the hardware aspects are dominant in the former and the software aspects predominate in the latter.

The node of this group about which there was most controversy was node 2, Computer Systems. Reviewers of the Committee's work have criticized it as a "misconceived hybrid," and its inclusion as an indication that the *Taxonomy* is typical committee product lacking in conceptual unity. None of the proposed alternatives, however, seemed to be an improvement, particularly in terms of creating a document understandable to the broad audience addressed. As stated earlier, the possible ephemeral nature of much of the material under this node, due to the rapid growth of VLSI technology, has been recognized.

But in any case, the notion of computer *systems* seems so important that to relegate it to any other than a first level node would be to do severe damage to the perspective on CS&E reflected in the *Taxonomy*.

Problems of another kind are illustrated in connection with node 3, Data. The reader may be surprised to learn that for a considerable time in the development of this *Taxonomy* there were only eight first level nodes, the one for Data having been inserted long after all the others. In retrospect, this seems strange to the Committee too, for while the early history of computers exhibited a focus on process or control rather than on data, that imbalance has been corrected for some time now. Witness, for example, the Database Management Systems node (7.4). Still it was believed for some time that most of the topics now appearing under the Data node could be appropriately included elsewhere (e.g. Software). But at one point the need for such a node became, it seemed, obvious to the Committee.

Following the four "core" nodes are two which may be characterized as the "theoretical" part of CS&E, Mathematics of Computing and Theory of Computation.

It has already been mentioned that node 6, Theory of Computation, is one where structure may be expected to change as the field develops. But node 5 also, Mathematics of Computing, may be expected to change for different reasons. The changing role of mathematics in computer science and engineering—more reliance on mathematics but less belief that certain types of mathematics (e.g. numerical analysis) are part of CS&E—may dictate change in future versions of this document.

In fact, a point worth discussing is why there should be a separate node 5 at all. It may well be that, as CS&E develops and matures, some or all of the topics under the Mathematics of Computing may migrate to node 7, Methodologies. For the present, however, it seems that so many computer scientists have worked and still work on the topics under node 5 that its centrality to the discipline needs to be recognized in the form of a separate node.

The final three nodes, Methodologies, Applications, and Milieux, all represent in a certain sense the "applied" side of CS&E (not "applied" in the sense of "implementation" of hardware and software, but in the sense of use for human and organizational purposes). Indeed, some would group all these topics under the blanket term "applications." The tripartite classification is useful, however, for distinguishing (1) topics representing broad classes of information processing techniques applicable in a variety of subject-matter contexts or disciplines; (2) topics representing these subject-matter contexts or disciplines themselves; and (3) the environment in which such subject-matter contexts or disciplines occur or are pursued.

The term "methodologies" for (1) was apparently first employed in the ACM Curriculum 68 report, precisely to distinguish such techniques from "applications" in the subject-matter sense. The term "milieux" for (3) has been used for some time in the category list for *Computing Reviews*.

Perhaps the most difficult task in developing this *Taxonomy* was to decide what to include—and exclude—from these three areas, and to decide which subtopics appropriately belonged in which. Indeed, extensive argument has occurred over whether (2) and (3) are appropriate taxonomic fodder at all. It is, therefore, worthwhile to discuss at some greater length the rationale behind the Committee's decisions on these nodes.

Looking at the list of subtopics under node 7, Methodologies, shows them all indeed to be areas which are studied and used in CS&E, and characterized essentially with respect to the form of information and techniques of processing it, usually without regard to "meaning" in the sense of real-world modeling. Thus the grouping of these subtopics under a separate heading from "applications" tends to encourage the view that, say, pattern recognition may be useful for other than "scientific" purposes, and database management systems are not only useful in "business" contexts.

Nevertheless, it must be expected that the particular list of subtopics here is perhaps the most volatile of all the major categories. Additions to this node can certainly be anticipated. Some may result from the proliferation of techniques used now for only one or a few applications. Still others may result from the fissioning of current methodologies into distinct categories (e.g., several years ago computer graphics and image processing would not have been distinct; neither would artificial intelligence and pattern recognition). And there is the possibility of migration into this node or migration from this node if subtopics in it assume more fundamental importance in the discipline.

In particular, considerable opinion was expressed that Algebraic Manipulation (7.1) is mathematics in the same sense as the topics under node 5 and, therefore, should be included there. On balance, however, it was decided that, while the importance of the topics under node 7.1 is growing, they still do not have a central enough place in the discipline to be considered more than a methodology. And in any case, rather than migrating to node 5, it seems equally likely, as noted above, that some or all of node 5 may migrate to node 7.

While the list of Methodologies is intended, at least in principle, to be comprehensive, a different approach is necessarily adopted for node 8, Applications. Here only a sampling of rather extensive categories are listed, without attempting to be complete. Even though the intent of node 8 is to be illustrative, however, it may well be that better illustrations will suggest themselves in the future and that it will seem desirable to expand this node modestly.

Because node 9 is concerned with the interface between CS&E and the world outside the discipline, it can be expected to undergo change as the impact of CS&E on life and society (e.g. by the rapid proliferation of personal computers) becomes increasingly profound. From a taxonomic point of view, however, what these changes are is less important than the desirability of retaining this node as a means to give the general reader perspective about CS&E.

The foregoing discussion has focused on the broad classification of the first level nodes, and of some issues bearing on seeming conflicts among them. Now some brief notes are given on each highest level node, regarding more detailed questions which arose concerning the subtopics included within it.

1. Hardware

The assigning and structuring of subtopics under hardware seemed mainly straightforward. Two particular questions were resolved as follows:

1. Do hybrid computers belong under Hardware or Computer Systems? Both are plausible, so much so that they appear under both (1.1.3 and 2.1.1) with different emphases.

2. How about microprogramming which has both hardware and software (micro*programming*) aspects? On balance it was decided that, quite aside from the programming activity of microprogramming, its crucial aspect is its effect on the control unit hardware (1.2.1.2).

2. Computer Systems

A number of people were bothered by the rather artificial — and to some degree invented — terms for nodes 2.1, 2.2 and 2.3. Uniquely (or almost so) among the first or second level nodes, the headings for these nodes are not common terms in computer science and engineering. They are, rather, "covering" nodes intended only to serve to group the nodes under them. This should perhaps be regarded as representing nothing more than the relatively undeveloped state of computer systems as a topic, as contrasted, say, with hardware/software.

3. Data

It has already been mentioned that the Data node came into being rather late. When it was decided to include such a node, the topics to be included under it became, with one exception, obvious.

The exception was the subtopic of data communications, and this illustrates how additional problems in achieving conceptual unity are caused by topics at the boundaries of CS&E. Some aspects of data communications are, the Committee is convinced, part of CS&E. Some would argue that data communications are so important in modern computer systems that this subject should be a first level node. But it was felt that this would overemphasize the position of data communications in CS&E. Another suggestion was to include it under node 2, Computer Systems, since the major role of data communications is in computer systems. The subject matter under node 2, however, seems very different in kind from that which should be covered under data communications.

The solution adopted, namely to place data communications at the second level under node 3, Data, may be superficially obvious, perhaps, but is not entirely satisfactory either. There is a static aspect to the other topics under node 3 which clashes with the dynamic, processing-related subject matter of data communications. Still, the essence of data communications is the data, and so there the topic resides.

4. Software

A seeming difficulty with this node is that software is ubiquitous, and is certainly an inextricable part of the subtopics under Computer Systems and Data. For example, data types and structures are a vital facet of programming languages, but the name itself implies an affinity with the Data node. They receive, therefore, their main treatment under that node (3.1 and 3.2) but are listed also—with appropriate cross-references—under Software (4.1.1.5).

5. Mathematics of Computing

Some of the fundamental areas of mathematics to which CS&E has contributed—and in which it has influenced directions of research—include automata theory, recursive functions, numerical analysis, combinatorics, graph theory, number theory, queueing theory, and arithmetic systems.

A most difficult problem facing the Committee was to decide what parts of mathematics to include and what parts to exclude. For example, number theory was excluded because, while the existence of computers has profoundly affected research in number theory, number theory itself is little used by and has been almost unaffected by computer scientists and engineers. And, similarly, it was decided to exclude almost all of probability and statistics, with the exception of random number generators and Monte Carlo methods, even though computers are used heavily to carry out statistical calculations and even though the development of computer-based statistical packages has been an active area of research. In general, an attempt has been made to include those areas of mathematics which focus on computational properties of algorithms and to which scientists who consider themselves computer scientists have made major contributions.

Even so, serious problems arise. Consider for instance the node on differential equations (5.1.3). All numerical methods for finding solutions of differential equations should surely be included as should the area of the numerical stability of an algorithm. On the other hand, questions regarding convergence properties are more difficult to decide. In general, convergence is a mathematical property and so should be excluded; however, when convergence is affected by roundoff error, it becomes also a computational property and for this reason should be included.

Some other aspects of the classification of first level nodes deserve mention. Node 5.1, Continuous Mathematics, includes those problems and areas which on physical and mathematical grounds are continuous in nature, even though methods used to solve them are discrete. Node 5.2, Discrete Mathematics, includes those problems and areas which on mathematical grounds are discrete or finite in nature, even though some of the methods used to solve them are continuous. Thus, for example, a linear system of equations has a discrete solution but iterative methods for solving them are continuous in nature since in general an infinite number of iterations is required for convergence. Node 5.3, Numerical Software and Algorithm Analysis, although it includes non-mathematical subnodes, is included here because of its importance in the development of mathematical software.

Finally, it is worth noting that inclusion of a topic in node 5 is not necessarily intended to imply that this subject matter has now become part of CS&E and is, therefore, no longer part of mathematics. For example, Graph Theory (5.2.2) and Combinatorics (5.2.3) are classical branches of mathematics and, while computer scientists have made and do make significant contributions to these disciplines, most research in these areas is still done by mathematicians.

6. Theory of Computation

While the topics under node 6 could all—or almost all—be classified as mathematics, as suggested earlier, the topics under node 6 seem so distinct in subject matter and outlook from those under node 5 that to include them under the same node would be quite misleading. Moreover, it seems probable that, whatever the future path of node 5, the theory of computation will always be central to CS&E and deserving of a separate node.

7. Methodologies

If, as in the definition of CS&E, a methodology refers to a method or technique of wide applicability, then the premise must be accepted that yesterday's application may be today's methodology. And, as already indicated, methodologies and core areas of the discipline may assume greater or lesser importance as the discipline develops, expands or contracts. Still, how do we justify the nine topics chosen under node 7 as methodologies?

First, the Committee is reasonably confident that the list is not too short. It is rather difficult to suggest areas of CS&E which meet the test of wide applicability which are not covered under node 7 (or, perhaps, in nodes 1-6; see, for example, the section on Mathematics of Computing). All suggestions for extension of the list of topics have failed to meet the test of wide applicability (e.g. computer-aided design) or have failed to be a bona fide part of CS&E (e.g. operations research).

Is the list perhaps too long? At least 6 of the nine second level nodes (Artifical Intelligence, Information Storage and Retrieval, Database Management Systems, Pattern Recognition, Modeling and Simulation and Computer Graphics) are quite widely taught as the subjects of university courses, a fact that suggests they are more than mere applications. Of these six, one—Artificial Intelligence (AI)—would be adjudged core computer science and engineering by some but the Committee believes it is prudent to treat AI as a methodology while admitting that future progress may well change this judgment. Another—Database Management Systems—is rapidly becoming so important that a change in categorization in the future might occur even though the discipline still needs to mature. A third—Modeling and Simulation—has many of the disciplinary characteristics of operations research but the Committee believes that the growth in importance of simulation has been so closely tied to computers that it, too, is a bona fide part of computer science.

Algebraic Manipulation is a borderline case. It has been extant for a considerable time and its growth in importance has been slower than in most other areas under node 7. Yet that growth has been steady and the most important algebraic manipulation systems are now quite widely used. Image Processing is also borderline. On the one hand, it is relatively new; on the other, it has considerable subject matter in common with Computer Graphics. But its growth is so rapid—and the applications so varied and different from "classical" computer graphics—that it seems to warrant separate treatment.

Sorting and Searching is rather different in character from the other second level nodes under node 7. Its subject matter is narrower than that of the other nodes and, while among the oldest of all computer techniques areas, it still retains some of the character of an application. Still, the uses are so common, so important and—primarily—so varied in character today that singling it out as a methodology seems justified.

8. Applications/Techniques

To have included all applications of CS&E in this *Taxonomy* would, inevitably, have resulted in an Applications node which dwarfed the others. Moreover, this would have resulted in little extra insight into the discipline of CS&E. Instead, the Committee has chosen to list under broad headings the applications that would, in effect, disappear if the computer disappeared. Thus, for example, engineers would still design bridges—and carry out the calculations necessary for so doing —even if there were no computers. In contrast, text processing would hardly exist without computers.

Notwithstanding this restriction, the list is far from inclusive. Many applications that simply would not exist in any recognizable form without the computer have not been mentioned (e.g. those associated with home computers, newly developed digital communication networks, military command and control systems). Instead the Committee has chosen to delineate in detail a few areas—four to be exact—that characterize applications made possible by the computer and that are so pervasive that they epitomize them.

Even with this approach, this node presents considerable difficulties. Is CAI (node 8.3) an application or is it a methodology with wide applicability? The Committee concluded that the level of success achieved by CAI thus far does not merit the methodology label, although this opinion might have to be modified in the future. Similarly, Text (Word) Processing (node 8.4) may well be so pervasive in the future as to merit being called a methodology, but it is not now.

9. Computing Milieux

Perhaps more than in most disciplines, complete understanding of computer science and engineering requires consideration of the economic, educational, legal, organizational and social milieux related to computing. At the risk of appearing chauvanistic, it seems to us that no discipline is having—or likely to have in the foreseeable future —as great an impact on the entire social fabric as CS&E. The Committee believes, therefore, that the topics covered in node 9 provide a valuable—even necessary—adjunct to the more typical taxonomic subjects covered by the other nodes.

VIII. REPRESENTATIVE BIBLIOGRAPHY

The numbers below refer to first-level (nodes 1, 2, 3, 4, 5, and 6) or second-level (nodes 7, 8, and 9) nodes on the TAXONOMY tree. This bibliography consists of books (except in a few special cases) concerned directly with the subject matter covered under the nodes.

1. Bell, C. G. and A. Newell (1971): COMPUTER STRUCTURES: READINGS AND EXAMPLES, McGraw-Hill.

Hamacher, V. C., Z. G. Vranesic and S. G. Zaky (1978): COMPUTER ORGANIZATION, McGraw-Hill.

Hayes, J.P. (1978): COMPUTER ARCHITECTURE AND ORGANIZATION, McGraw-Hill.

Mead, C. and L. Conway (1980): INTRODUCTION TO VLSI SYSTEMS, Addison-Wesley.

Williams, G. E. (1977): DIGITAL TECHNOLOGY, Science Research Associates.

2. Enslow, Philip H. (ed.) (1974): MULTIPROCESSORS AND PARALLEL PROCESSING, John Wiley & Sons.

Ferrari, D. (1978): COMPUTER SYSTEMS PERFORMANCE EVALUATION, Prentice-Hall.

Martin, James (1967): DESIGN OF REAL TIME COMPUTING SYSTEMS, Prentice-Hall.

Stone, Harold S. (1975): INTRODUCTION TO COMPUTER ARCHITECTURE, SRA.

3. Abramson, N. (1963): INFORMATION THEORY AND CODING, McGraw-Hill.

Davis, D. W. and D. L. A. Barber (1973): COMMUNICATION NETWORKS FOR COMPUTERS, John Wiley & Sons.

Horowitz, E. and S. Sahni (1976): FUNDAMENTALS OF DATA STRUCTURES, Computer Science Press.

Knuth, D. E. (1973): THE ART OF COMPUTER PROGRAMMING (vol. 1): FUNDAMENTAL ALGORITHMS (2nd ed.), Addison-Wesley.

McNamara, J. E. (1977): TECHNICAL ASPECTS OF DATA COMMUNICATION, Digital Equipment Corporation.

Tremblay, J. P and P. G. Sorenson (1976): AN INTRODUCTION TO DATA STRUCTURES WITH APPLICATIONS, McGraw-Hill.

4. Graham, R. M. (1975): PRINCIPLES OF SYSTEMS PROGRAMMING, John Wiley & Sons.

Gotlieb, C. C. and L. R. Gotlieb (1978): DATA TYPES AND STRUCTURES, Prentice-Hall.

Linger, R. C., H. D. Mills and B. I. Witt (1979): STRUCTURED PROGRAMMING: THEORY AND PRACTICE, Addison-Wesley.

Nicholls, J. E. (1975): THE STRUCTURE AND DESIGN OF PROGRAMMING LANGUAGES, Addison-Wesley.

Sammet, J. E. (1969): PROGRAMMING LANGUAGES: HISTORY AND FUNDAMENTALS, Prentice-Hall.

Tsichritzis, D. C. and P. A. Bernstein (1974): OPERATING SYSTEMS, Academic Press.

Weinberg, G. (1971): THE PSYCHOLOGY OF COMPUTER PROGRAMMING, Van Nostrand Reinhold.

Yourdon, E. (1975): TECHNIQUES OF PROGRAM STRUCTURE AND DESIGN, Prentice-Hall.

5. Ames. W. F. (1978): NUMERICAL METHODS FOR PARTIAL DIFFERENTIAL EQUATIONS, Academic Press.

Conte, S. D. and Carl de Boor (1971): ELEMENTARY NUMERICAL ANALYSIS, McGraw-Hill.

Freiberger, W. and U. Grenander (1971): A SHORT COURSE IN COMPUTATIONAL PROBABILITY AND STATISTICS, Springer-Verlag.

Kleinrock, L. (1975): QUEUEING SYSTEMS, John Wiley & Sons.

Percus, J. K. (1971): COMBINATORIAL METHODS, Springer-Verlag.

Ralston, A. and P. Rabinowitz (1978): A FIRST COURSE IN NUMERICAL ANALYSIS (2nd ed.), McGraw-Hill.

6. Aho, A. V., J. E. Hopcroft and J. D. Ullman (1974): THE DESIGN AND ANALYSIS OF COMPUTER ALGORITHMS, Addison-Wesley.

Borodin, A. and I. Munro (1975): THE COMPUTATIONAL COMPLEXITY OF ALGEBRAIC AND NUMERIC PROBLEMS, Elsevier-North Holland.

Hopcroft, J. E. and J. D. Ullman (1969): FORMAL LANGUAGES AND THEIR RELATION TO AUTOMATA, Addison-Wesley.

Horowitz, E. and S. Sahni (1978): FUNDAMENTALS OF COMPUTER ALGORITHMS, Computer Science Press.

Manna, Z. (1976): MATHEMATICAL THEORY OF COMPUTATION, McGraw-Hill.

Rustin, R. (ed.) (1973): COMPUTATIONAL COMPLEXITY, Algorithmic Press.

7.1 Knuth, D. E. (1969): THE ART OF COMPUTER PROGRAMMING (vol. 2): SEMI-NUMERICAL ALGORITHMS, Addison-Wesley.

Petrick, S. (ed.) (1971): PROCEEDINGS OF THE SECOND SYMPOSIUM ON SYMBOLIC AND ALGEBRAIC MANIPULATION, ACM.

Jenks, R. D. (ed.) (1976): PROCEEDINGS OF THE 1976 SYMPOSIUM ON SYMBOLIC AND ALGEBRAIC COMPUTATION, ACM.

7.2 Hunt, Earl B. (1975): ARTIFICIAL INTELLIGENCE, Academic Press.

Jackson, Philip C.(1974): INTRODUCTION TO ARTIFICIAL INTELLIGENCE, Van Nostrand Reinhold.

Raphael, Bertram (1976): THE THINKING COMPUTER: MIND INSIDE MATTER, W. H. Freeman and Company.

Winston, Patrick H. (1977): ARTIFICIAL INTELLIGENCE, Addison-Wesley.

7.3 Lancaster, F. W. and E. G. Gayen (1973): INFORMATION RETRIEVAL ON-LINE, Melville.

Salton, G. (1975): DYNAMIC INFORMATION AND LIBRARY PROCESSING, Prentice-Hall.

Van Rijsbergen, C. J. (1975): INFORMATION RETRIEVAL, Butterworths.

7.4 ACM Computing Surveys, SPECIAL ISSUE ON DATA BASE MANAGEMENT SYSTEMS, vol. 8, March 1976.

Date, C. J. (1977): AN INTRODUCTION TO DATA BASE SYSTEMS (2nd ed.), Addison-Wesley.

Tsichritzis, D. and F. Lochovsky (1977): DATA BASE MANAGEMENT SYSTEMS, Academic Press.

7.5 Gonzales, R. C. and P. Wintz (1977): DIGITAL IMAGE PROCESSING, Addison-Wesley.

Pratt, W. K. (1978): DIGITAL IMAGE PROCESSING, John Wiley & Sons.

Rosenfeld, A. and A. C. Kak (1976): DIGITAL PICTURE PROCESSING, Academic Press.

7.6 Batchelor, B. G. (ed.) (1978): PATTERN RECOGNITION - IDEAS IN PRACTICE, Plenum Press.

Chen, C. H. (1973): STATISTICAL PATTERN RECOGNITION, Hayden.

Ullman, J. R. (1973): PATTERN RECOGNITION TECHNIQUES, Butterworths.

7.7 Gordon, G. (1978): SYSTEM SIMULATION (2nd. ed.), Prentice-Hall.

Roth, P. F. (1976): "Simulation" in ENCYCLOPEDIA OF COMPUTER SCIENCE, (A. Ralston, ed.), Van Nostrand Reinhold.

Shannon, R. E. (1975): SYSTEMS SIMULATION, THE ART AND SCIENCE, Prentice-Hall.

Zeigler, B. P., M. F. Elzas, G. J. Klir, and T. I. Oren (eds.) (1979): METHODOLOGY IN SYSTEMS MODELLING AND SIMULATION, Elsevier-North Holland.

7.8 Knuth, D. E. (1973): THE ART OF COMPUTER PROGRAMMING (vol. 3): SORTING AND SEARCHING, Addison-Wesley.

Lorin, Harold (1975): SORTING AND SORT SYSTEMS, Addison-Wesley.

Rich, R. P. (1972): INTERNAL SORTING METHODS ILLUSTRATED WITH PL/1 PROGRAMS, Prentice-Hall.

7.9 Giloi, W. K (1978): INTERACTIVE COMPUTER GRAPHICS, Prentice-Hall.

Newman, W. and R. Sproull (1978): PRINCIPLES OF INTERACTIVE COMPUTER GRAPHICS, McGraw-Hill.

Prince, D. (1970): INTERACTIVE GRAPHICS FOR CAD, Addison-Wesley.

8.1 Dolotta, T. A. et al (1976): DATA PROCESSING IN 1980-1985, John Wiley & Sons.

McLean, E. R. and V. V. Soden (1977): STRATEGIC PLANNING FOR MIS, John Wiley & Sons.

Parkhill, D. F. (1976): THE CHALLENGE OF THE COMPUTER UTILITY, Addison-Wesley.

8.2 **Bovington, P. R.** (1969): DATA REDUCTION AND ERROR ANALYSIS FOR THE PHYSICAL SCIENCES, McGraw-Hill.

Chambers, J. M. (1977): COMPUTATIONAL METHODS FOR DATA ANALYSIS, John Wiley & Sons.

Daniel, C. and F. S. Wood (1971): FITTING EQUATIONS TO DATA, John Wiley & Sons.

Ehrenberg, A. S. C. (1975): DATA REDUCTION: ANALYZING AND INTERPRETING STATISTICAL DATA, John Wiley & Sons.

8.3 **Lecarme, O. and R. Lewis** (eds.) (1975): COMPUTERS IN EDUCATION (2 vols.), Elsevier-North Holland.

Levien, R. E. (1972): THE EMERGING TECHNOLOGY: INSTRUCTIONAL USES OF THE COMPUTER IN HIGHER EDUCATION, McGraw-Hill.

8.4 **Van Dam, A. and D. Rice** (1971): "On-line Text Editing: A Survey," COMPUTING SURVEYS, vol. 3, no. 3. pp. 93-114. See also article on Text Editing in ENCYCLOPEDIA OF COMPUTER SCIENCE (1976), A. Ralston (ed.), Van Nostrand Reinhold.

9.1 **Phister, M.** (1975): DATA PROCESSING TECHNOLOGY AND ECONOMICS (2nd. ed.), Santa Monica Publishing Company.

(1973): THE STATE OF THE COMPUTER INDUSTRY IN THE UNITED STATES, AFIPS Press.

(1976): THE AMERICAN COMPUTER INDUSTRY IN ITS INTERNATIONAL COMPETITIVE ENVIRONMENT, U.S. Department of Commerce.

9.2 **Austing, R. H., B. H. Barnes and G. I. Engel** (1977): "A Survey of Literature in Computer Science Education," COMMUNICATIONS OF THE ACM, vol. 20, pp. 13-21.

Bukeski, W. J. and A. L. Korotkin (1975): COMPUTING ACTIVITIES IN SECONDARY EDUCATION, American Institute for Research.

Lecarme, O. and R. Lewis (eds.) (1975): COMPUTERS IN EDUCATION (2 vols.), Elsevier-North Holland.

9.3 **Goldstine, H. H.** (1972): THE COMPUTER FROM PASCAL TO VON NEUMANN, Princeton University Press.

Randell, B. (1973): THE ORIGINS OF DIGITAL COMPUTERS: SELECTED PAPERS, Springer-Verlag.

Wexelblat, R. (ed.) (1980): FINAL PROCEEDINGS OF ACM SIGPLAN HISTORY OF PROGRAMMING LANGUAGES CONFERENCE, Academic Press.

Metropolis, N., G.C. Rota and J. Howlett (1980): PROCEEDINGS OF LOS ALAMOS HISTORY CONFERENCE, Academic Press.

9.4 **Bernacchi, R. and G. Larsen** (1974): DATA PROCESSING CONTRACTS AND THE LAW, Little Brown.

Bigelow, R. (1972): COMPUTER LAW SERVICE, Callaghan Co. (looseleaf—updated).

Bigelow, R. and S. Nycum (1976): YOUR COMPUTER AND THE LAW, Prentice-Hall.

9.5 **Brooks, F. P. Jr.** (1975): THE MYTHICAL MAN-MONTH: ESSAYS ON SOFTWARE ENGINEERING, Addison-Wesley.

Dolotta, T. A. et al (1976): DATA PROCESSING IN 1980-1985, John Wiley & Sons.

Orlicky, J. (1969): THE SUCCESSFUL COMPUTER SYSTEM, McGraw-Hill.

9.6 See articles on Societies and Organizations in ENCYCLOPEDIA OF COMPUTER SCIENCE (1976), A. Ralston (ed.) Van Nostrand Reinhold.

9.7 **Gotlieb, C. C. and A. Borodin** (1973): SOCIAL ISSUES IN COMPUTING, Academic Press.

Mowshowitz, A. (1976): THE CONQUEST OF WILL: INFORMATION PROCESSING IN HUMAN AFFAIRS, Addison-Wesley.

Weizenbaum, J. (1976): COMPUTER POWER AND HUMAN REASON, W. H. Freeman.

IX. ABBREVIATIONS AND ACRONYMS

ACM	Association for Computing Machinery
ADAPSO	Association of Data Processing Service Organizations
AFIPS	American Federation of Information Processing Societies
AI	Artificial Intelligence
ANSI	American National Standards Institute
ARPA	Advanced Research Projects Agency (of the Department of Defense; now D(efense)ARPA)
ASCII	American Standard Code for Information Interchange
AVL	Adel'son-Vel'skii-Landis (after two Russian mathematicians)
BNF	Backus-Naur Form
CAI	Computer-Assisted Instruction
CBEMA	Computer and Business Equipment Manufacturers Association
CBCT	Customer Bank Communications Terminal
CCITT	Comité Consultatif International Télégraphique et Téléphonique
CCP	Certificate in Computer Programming
CDC	Control Data Corporation
CDP	Certificate in Data Processing
CMOS	Complementary Metal-Oxide Semiconductor
COM	Computer Output on Microfilm
CPU	Central Processing Unit
CRT	Cathode Ray Tube
CS&E	Computer Science and Engineering
CTSS	Compatible Time-sharing System
DEC	Digital Equipment Corporation
DECUS	Digital Equipment Corporation Users Society
DPMA	Data Processing Management Association
EBCDIC	Extended Binary-Coded Decimal Interchange Code
EDSAC	Electronic Delay Storage Automatic Calculator
EDVAC	Electronic Discrete Variable Automatic Calculator
EEG	Electroencephalogram
EFT	Electronic Funds Transfer
EIA	Electronic Industries Association
EKG	Electrocardiogram
ENIAC	Electronic Numerical Integrator And Computer
FAP	Fortran Assembly Program
FCC	Federal Communications Commission
FIFO	First-In-First-Out
IBM	International Business Machines Corp.
ICCP	Institute for Certification of Computer Professionals
IEEE	Institute for Electrical and Electronic Engineers
ILLIAC	Illinois Automatic Computer
I/O	Input/Output
ISO	International Standards Organization

KWIC	Keyword-In-Context	**SDI**	Selective Dissemination of Information
LIFO	Last-In-First-Out		
MICR	Magnetic Ink Character Recognition	**SOAP**	Symbolic Optimizer and Assembly Program
MIT	Massachusetts Institute of Technology	**SSEC**	Selective Sequence Electronic Calculator
NCR	National Cash Register Company	**TRANSAC**	Transistor Automatic Computer
NMOS	N-channel Metal-Oxide Semiconductor	**TV**	Television
NMR	Nuclear Magnetic Resonance	**UNIVAC**	Universal Automatic Computer
OCR	Optical Character Recognition	**USE**	UNIVAC Scientific Exchange
OS	Operating System	**VDL**	Vienna Definition Language
PDP	Programmed Data Processor		
PMOS	P-channel Metal-Oxide Semiconductor		
RCA	Radio Corporation of America		
SAP	Symbolic Assembly Program		

X. INDEX

Key

(c) A core term

(u) An umbrella term

1.5.1.2 Labels a term occurring as all or part of a node name. Its annotation usually defines the term.

6.1.4.5 Labels a term cited in the annotation to node 6.1.4.5. The term may or may not be defined there; it is usually only mentioned in context.

Abbreviations used to clarify the context of certain terms cited in the index:

AI Artificial Intelligence

CAI Computer-assisted Instruction

CG Computer Graphics

DBMS Database Management System

IP Image Processing

IR Information Retrieval

PR Pattern Recognition

Abacus	**9.3.1.1.1**
Absolute address	**3.2.1.3.1** 4.1.1.3.2 4.2.2.2.1
Absolute value	5.1.6.2.4 7.1.2.2.4
Abstract data type	**4.1.1.5.3**
Abstract language families	**6.2.7.2**
Abstract syntax	**6.3.1.2**
Abstracting procedures (IR)	**7.3.1.2**
Acceptance testing	**9.5.2.2.6**
Acceptor	**6.1.3** 6.2.1.1
Access-based computer systems	**2.2**
Access methods (file)	**4.3.1.2.5** 3.2.3.2 7.4.3 7.4.5.2.1
Access time	1.2.2.2.1.2
Accounting	**8.1.1.1**
Accuracy	5.2.1.3 5.3.1.3
ACM	**9.6.2.1.2**
Acoustic coupler modem	1.4.5.2 3.4.1.2.2
Acoustic waveform	7.2.4.1
Acyclic graph	**3.1.2.6.1.1**
ADAPSO	**9.6.2.3.1**
Adaptive coding (IP)	**7.5.2.4**
Adaptive control	1.1.3.2 **2.3.2**
Adaptive quadrature	**5.1.2.6**
Adder circuit	**1.5.1.4.5**
Address (c)	1.2.2.2.2
Address calculation sort	**7.8.1.1.6.2**
Advanced development	**9.5.3**
AFIPS	**9.6.2.1.1**
Aiken, Howard	9.3.2.2.4
Air traffic control	2.3.4.1 4.2.1.2.1
Algebraic functions	**7.1.1.2.3**
Algebraic manipulation	**7.1** 8.2.4.2
Algebraic substitution cipher	**3.3.4.3**
Algol	See delimiter (c)
Algorithm (c)	**5.3** 7.1.2 3.3.4.2 5.2.3.1.4 See also analysis of algorithms
Algorithm logics	**6.3.4.4**
Aliasing (CG)	**7.9.7.4**
Alphanumeric (c)	
Alternating direction implicit (ADI) method	**5.1.4.1.4**
Ambiguous grammar	**6.2.3.2**
Amdahl 470 V/6	9.3.3.3.1.3
Analog circuitry	**1.5.2**
Analog computer	**1.1.2** 8.2.1.1.1
Analog switching system	3.4.2.1.1
Analog-to-digital (A to D) converter	**1.5.3.1**
Analysis of algorithms	**5.3** 6.6 7.1.2.7 7.1.4.3.1 7.8.5
Analysis of experimental data	**8.2.2**
Analysis of variance	**8.2.2.3.1**
Analytic complexity	**6.6.1.5**
Analytic engine	9.3.1.3.2
Analytic model of a system	**2.4.2.1.1.2**
Animation (CG)	**7.9.7.6** 9.7.7.3
Antitrust	**9.4.7**
Applications	**8.**

INDEX

Applications programming (u)	**9.1.3.2.2 9.5.5.4.3** 7.4.5.5
Application-based computer systems	**2.3.4**
Approximation	**5.1.6.2**
Apt	See standards (u)
Architecture, computer	See computer architecture
Argument (c)	
Arithmetic data	**4.1.1.5.1.1 3.1.1.1-2**
Arithmetic-logic unit	**1.2.1.1**
Arithmetic register	See register (c)
ARPA net	**9.3.3.3.3.1**
Array	**4.1.1.5.2.1 3.1.2.2**
	See also loop (c)
Array processor	**1.3.2.1.3 2.1.3.2 9.3.3.3.3.2**
Artificial intelligence (AI)	**7.2** See also cybernetics (u)
Arts, computing in the	**9.7.7.3** 8.3.3.7
ASCII code	**3.2.1.2.2** See also standards (u)
Assembly language	**4.1.1.3.2 9.3.3.1.4.1**
	See also object program (c)
Asset management	**8.1.1.3**
Assignment statement	**4.1.1.4.1.1 6.3.3.1**
	See also expression (c)
Associative memory/processor	**1.2.2.2.2 1.3.2.1.1**
	7.4.5.10.3 3.1.2.4.1
Astronomy, algebraic manipulation in	**7.1.4.4**
Asynchronous processing	2.1.3.3 2.3.4.6
	3.4.1.3.3.2 6.3.3.2.2.2
Atanasoff, John	**9.3.2.3.2**
Atlas computer	**9.3.3.2.3**
Attenuation errors	**3.4.1.4.1.2**
Attributes	**7.7.2.2.4**
Audio I/O	**8.3.1.3.4-5**
	See also speech recog. & response
Auditing	**9.5.1.4**
Augmented transition network	**7.3.1.4.3**
Authoring (CAI)	**8.3.4**
Automata theory	**6.1 6.1.4**
Automated teller	**9.4.8.3.1**
Automatic control theory	See cybernetics (u)
Automatic feedback (IR)	**7.3.3.4.4**
Automatic index generation	**8.4.3.3**
Automatic programming	**4.1.1.1.4**
Automatic typesetting	7.3.6.2.3
Automation (u)	
Auxiliary information systems	7.3.6.2 **8.4.3**
Auxiliary storage	**7.4** 7.4.5.2.3
Available storage list	**3.3.1.6**
AVL tree	**3.1.2.5.1.2** 3.2.3.1.4
Axiom	**7.2.2.2**
Axiomatic semantics	**6.3.2.3**
Babbage, Charles	**9.3.1.3**
Backend computer	**7.4.5.10.1**
Backtracking	**7.2.10.1.1.2** 6.2.4.2
Backus-Naur form (BNF)	**4.1.1.1.2**

431

Balanced tape merge	**7.8.1.2.2.1**
Banking	**9.4.8.3**
Basic (language)	9.3.3.3.2.2
Batch processing	**2.2.1 4.2.1.2.3**
	7.1.3.2.2 9.5.3.4.3
Batcher's parallel sort	**7.8.1.1.3.2**
Bayesian statistical analysis	7.6.2.2.1 8.2.2.3.4
BCS	**9.6.2.1.5**
Bell Telephone Laboratory relay computer	**9.3.2.2.2**
Best-fit storage management	**3.3.1.3**
Biased exponent	**3.2.1.1.2.1**
Bibliometrics (IR)	**7.3.2.3.4**
Bilinear forms	**6.6.1.3**
Bimodal distribution	8.2.2.2
Binary (c)	5.2.1.5
Binary relations	6.3.2.2.3
Binary search	**7.8.2.2.1**
Binary tree	**3.1.2.5.1** 5.2.2.4.3
	7.8.1.1.1.3 7.8.1.1.2.4
Binomial coefficients	**5.3.1.3**
Bipolar semiconductor element	**1.6.1.1**
Bisection	**5.1.1.1**
Bit (c)	
Bit slice	1.2.2.2.1.3
Block design	5.2.3.6
Block-oriented simulation language	7.7.2.3.2.2
Boolean algebra	**6.1.1.1** 6.6.3.2.1
	7.3.3.1.3 6.3.4.4
Boundary value problem	**5.1.3.7** 5.1.3.7.4
	5.1.3.7.6 5.1.3.7.8
Bounded context grammar	6.2.6.2
Boxing (CG)	**7.9.3.2.5**
Branch of a tree or graph	3.1.2.5 5.2.2
Branching (c)	6.4.2
Brunsviga calculator	9.3.1.4.2
Bubble sort	7.8.1.1.3.3
Bubble memory	**1.6.2.3**
Buddy system	**3.3.1.5**
Budget analysis	**8.1.6.1**
Buffer (c)	1.2.2.1.4 7.4.5.2.3 **7.8.3.5**
Bug (c)	9.5.5.4.2
Burroughs adding machine	**9.3.1.4.1**
Burroughs datatron	9.3.3.1.2
Burroughs 1700	See cell (c)
Burroughs 6700 & 7700	9.3.3.3.1.3
Bus	6.4.3.1.3
Bush, Vannevar	9.3.2.2.1
Business data processing	**8.1** 4.1.1.3.1
Byte (c)	1.2.2.2.1.3
B-tree	7.8.2.2.3
Cable television	9.4.6.2.2
Cache memory	**1.2.2.1.1**
Calculators (early)	**9.3.1.2 9.3.1.4**

INDEX

Calligraphic display	7.9.1.1.1.1
Call-by-name	6.3.3.1.2.1
Call-by-reference	6.3.3.1.2.1
Call-by-value	6.3.3.1.2.1
Canonical forms	**6.2.3.3 7.1.2.2.1**
Capacity management	**9.5.5.5**
Card	See hollerith card (c)
Card programmed calculator	9.3.2.1.2
Card reader/punch	**1.4.1.1**
Cartridge devices	**1.4.3.1.2 1.4.3.2.2**
Case grammar (IR)	**7.3.1.4.5**
Case statement	See structured programming (u)
Cash flow analysis	**8.1.6.3**
Cash management	**8.1.1.2**
Cassette tape device	**1.4.3.1.1**
CBEMA	**9.6.2.3.2**
CCP certification	**9.6.4.1**
CDC Star computer	9.3.3.3.3.3
CDC 1604	9.3.3.2.2
CDC 3000 series	9.3.3.2.2
CDC 6000/Cyber	9.3.3.3.1.3 See also cell (c)
CDP certification	**9.6.4.1.2**
Cell, memory (c)	1.2.2.1.2
Central processing unit (CPU)	**1.2.1 9.1.3.1.1** 1.1.1.1-2
	1.2.3.1.1 1.2.3.1.2.1-2
	1.2.3.2.2 2.1.2 4.2.1.3.5
Centralization/decentralization (DBMS)	**7.4.6.3**
Centroid of a tree	**5.2.2.4.2**
Certification	**5.3.2.3 9.2.1.5.3 9.6.4.1** 9.6.2.4
Chained file	**7.3.2.1.2**
Channel (c)	**1.2.3.1.2.2**
	3.4.1.3.3.1 6.4.3.1.3
Character (c)	**3.1.1.4 3.2.1.2**
	4.1.1.5.1.3 1.4.1.3 4.1.1
Character printer	**1.4.1.3.2**
Character string	**4.1.1.5.1.4 3.1.1.5**
	See also symbol manipulation (u)
Characteristics	5.1.4.2.3
Charge-coupled storage device	**1.6.3**
Chemical structure diagrams (PR)	**7.6.3.2.2**
Chemistry, AI applications to	**7.2.9.1**
Chief-programmer team	See structured programming (u)
Chip	1.1.1.1 1.6.3
	See also integrated circuit
Chomsky hierarchy	**6.2.1**
Chromatic number of a graph	**5.2.2.3.5**
Church's lambda calculus	6.3.2.1.3
Cipher	3.3.4.1-4
Circuit board	1.1.1.2
Circuit complexity	**6.5.3.4**
Circuit in a graph	**5.2.2.2** 5.2.2.4
Circuit switching	**3.4.2.1**
Citation indexing (IR)	**7.3.1.1.7**
Classifier design (PR)	**7.6.2.3**

Cliques of a graph	**6.6.3.2.2** 5.2.3.7
Clipping (CG)	**7.9.3.2.2 7.9.1.1.7.4**
Clock input	1.5.1.4.1
Closed subroutine	See subroutine (c)
Closure	**6.1.3.1 6.2.3.4**
Cluster analysis algorithms (PR)	**7.6.2.1.2**
Clustered file	**7.3.2.1.5**
Coarseness of a graph	**5.2.2.3.4**
Cobol	4.1.1.3.4 7.4.4.2.2 See also standards (u)
Code (c)	4.1.1.4.1.5
Coding (encoding) of data	3.3.2
Coefficient domains	7.1.1.4
Cofactor expansion of a matrix	5.2.4.6.1
Cogo	4.1.1.3.4 5.3.2.6
Coherence criteria (IR)	7.3.1.2.3
Collection growth (IR)	7.3.2.3.1
Collection retirement (IR)	7.3.2.3.2
Collocation	5.1.3.7.4
Colorability of a graph	5.2.2.3.5
Colossus	9.3.2.3.3
Combinational circuit	**1.5.1.1 6.1.1.2** 1.5.1.3 1.5.1.4.4
Combinations	5.2.3.1.2
Combinatorics	5.2.3 6.6.2 7.1.4.2.3
Combinatory logic models	6.3.2.1.4
Command	**4.1.1.4** See also field (c)
Command language editor	8.4.1.3.1
Command sequence program	6.3.2.2.1
Comments (program)	4.1.3.4.1.1
Common carriers	**9.1.3.4.2 9.4.6.2.2** 9.4.6.1 1.4.5
Communication path	1.2.3.1 2.3.4.5
Communications systems	**2.2.4 3.4 3.4.3.2 9.1.3.1.5 9.1.3.4 9.4.6 9.5.1.3.5 9.5.3.2.3 7.9.2.5 9.5.1.1.5**
Community college education	9.2.1.3
Compaction (compression)	3.3.3 7.5.2 7.4.5.2.2
Comparator circuit	1.5.1.4.4
Comparison statement	7.8.5.1
Compiler	**4.1.1.2.1.1 9.3.3.1.4.2** 4.1.4.4 5.3.2.2
Compiler generator	4.1.1.2.2.1
Complete partial order semantics	**6.3.2.2.5**
Complex number	3.1.2.2 5.1.1
Complexity of computation	**6.5 6.2.4.3**
Component	1.1 1.1.3
Component improvement	1.7.3.2
Composite data structure	3.1.2
Composition of text	8.4.2.3
Compression (compaction)	3.3.3 7.5.2 7.4.5.2.2
Comptometer	9.3.1.4.3
Computation, theory of	**6.**
Computational complexity	**6.5**

INDEX

Computational mathematics education	9.2.1.2.4
Computer architecture	1.3 7.4.5.10 5.3.1.5
Computer arithmetic	5.2.1
Computer-assisted instruction (CAI)	8.3 7.2.9.2 9.2.2.1
Computer-assisted medical diagnosis	7.2.9.3
Computer circuitry	1.5
Computer-controlled experiments	8.2.1.2
Computer elements	1.6
Computer engineer	9.6.1.2
Computer engineering education	9.2.1.1.3 9.2.1.2.3
Computer graphics (CG)	7.9
Computer industry	9.1
Computer literacy	9.2.1.2.7 9.2.1.3.4
Computer models	6.4
Computer music	9.7.7.3
Computer network	1.3.2.2.4
Computer operations	9.5.5.2
Computer operator	9.6.1.5
Computer organization	9.5.6
Computer science education	9.2.1.1.1 9.2.1.2.1
Computer science, algebraic manipulation in	7.1.4.3
Computer scientist	9.6.1.1
Computer systems	2.
Computer-aided design	1.7.5
Computing milieux	9.
Computing services	9.1.3.3 9.5.4.1
Computing technology	9.7.2.1-2
Concatenation (c)	6.1.3.3 6.2.2.2 6.2.3.4 6.6.4.1 6.6.5.2
Concentrators, I/O	2.3.3.1.2
Concept formation (AI)	7.2.3.1
Concept hierarchy (IR)	7.3.1.3.5
Conceptual schema (DBMS)	7.4.1.3
Concordance (IR)	7.3.1.3.6
Concrete syntax	6.3.1.1
Concurrency of access (DBMS)	7.4.5.3
Conditional statement	4.1.1.4.1.3 6.3.3.2.1.1
Conditioning, transmission line	3.4.1.4.2
Conditioning of a linear system	5.2.4.2
Configuration	9.5.5.5.2
Conjugate gradient iterative method	5.2.4.4.4
Connectivity of a graph	5.2.2.2.1 5.2.2.4.7
Console (c)	9.5.5.2.1
Constant	6.1.1.1
Construct, program	6.3.3
Consulting	9.5.4.2.1
Consumer credit	9.4.8.5
Contact circuit	1.5.1.1.1
Content access array	3.1.2.2.2
Content addressable memory	See associative memory
Content analysis (IR)	7.3
Contention	3.4.1.3.3.3
Context-free grammar	6.2.1.3
Context-free languages	6.2.4-5

Context-sensitive grammars	**6.2.1.2**
Context-sensitive languages	**6.2.6**
Continuing education in computer science	**9.2.1.5 9.2.2.3**
Continuous mathematical variable	**5.1** 1.1.2
Continuous system simulation model	**7.7.3**
Contracts	**9.4.5 9.4.1.5**
Control grammars	**6.2.7.1**
Control of program sequence	**6.3.3.2**
Control signaling	**3.4.2.1.3**
Control structures	**4.1.1.4.1** 7.2.10.2
	See also structured programming (u)
Control unit	**1.2.1.2**
Controlled vocabulary	**7.3.3.1.1**
Controllers	**8.2.1.2.1-2**
Contour (IP)	**7.5.2.1.3**
Convergence	**5.1.3.5 5.1.4.2.4 5.3.1.1**
Conversion between systems	**9.5.2.3.1**
Cook reducibility	**6.6.3.1.1**
Coordinate addressed memory access	**1.2.2.2.1**
Copyrights	**9.4.1.1** 9.4.3.2
Coroutine	**4.1.2.2 7.2.10.1.2.2**
Correction of software	**4.1.3.6.1**
Correlation coefficients (IR)	**7.3.3.3.1**
Cost center/profit center issues	**9.5.6.1.2**
Counter machine automata	**6.1.5.1**
Counter circuit	**1.5.1.4.3**
CPU	See central processing unit
CPU resident controller	**1.2.3.1.2.1**
Cramer's rule	**5.2.4.6**
Cray-1 computer	**9.3.3.3.3.3**
Cross compiler	**4.1.1.2.1.4**
Cross section of an image	**7.5.7.4**
Cross tabulation	**8.2.2.3.4**
Crossing number of a graph	**5.2.2.3.4**
Crosstalk	**3.4.1.4.1.3**
CRT terminal	**1.4.2.2 1.4.3.4.1**
	7.6.4.1.3 7.9.1.1.1
Cryogenic storage element	**1.6.4**
CTSS time sharing system	**9.3.3.3.2.2**
Current awareness systems (IR)	**7.3.6.1.3**
Cursor (CG)	**7.9.1.2.1 8.4.1.3.2**
Curve generator	**7.9.1.1.7.1**
Custom software	**9.1.3.2.3**
Cut set of a graph	**5.2.2.3.3 5.2.2.4.7**
Cybernetics (u)	
Cycle in a graph	**5.2.2.2**
Cycle time (c)	
Dartmouth Basic time sharing system	**9.3.3.3.2.2**
Data acquisition	**8.2.1**
Data analysis	**5.1.6.2.8**
Data auditing (DBMS)	**7.4.6.4**
Data communications	**3.4**
Data definition/description language	**4.1.1.3.8 4.3.2.1 7.4.4.1**

INDEX

Data dictionary/directory (DBMS)	**7.4.2.4**
Data encoding	**3.3.2**
Data encryption standard	3.3.4.4
Data entry/retrieval	**1.4.1** 2.3.4.6.4 3.4 **4.2.3.2** 9.1.3.1.4 **9.5.5.3.1** 9.2.3.1.2
Data flow	**1.3.2.2.2** 1.2.2.1.4
Data format	1.2.2.1.4
Data General Nova	9.3.3.3.1.5
Data independence (DBMS)	**7.4.2.6**
Data management	**4.3 7.4.6** 3.3
Data manipulation language (DBMS)	**7.4.4.2** 7.4
Data models	**4.3.2.3.2 7.4.2.1 7.4.1.4**
Data movements	**7.8.5.2**
Data path (c)	**9.4.6.2**
Data processing (u)	**8.1-2 9.2.1.3.2 9.2.1.4.3**
Data quality and integrity (DBMS)	**7.4.2.5**
Data reduction	**7.7.4.3 8.2.2.4**
Data semantics (DBMS)	**7.4.1.1**
Data smoothing	**5.1.6.2.8 8.2.2.1.1**
Data storage representation	**3.2**
Data structures	3.1 **4.1.1.5.2 4.3.2 7.1.1.1** 7.2.4.2 7.2.10.2 **7.8.3.1**
Data switching	**3.4.2**
Data tablet	7.9.1.2.1
Data transfer	**4.2.2.3** 1.2.3.2.2 7.8.4.3
Data transmission	**3.4.1**
Data types	**4.1.1.5.1**
Database administrator	4.3.2.3.1 **7.4.6.2 9.6.1.6**
Database machines	**7.4.5.10.2**
Database management system (DBMS)	**7.4 9.1.3.3.3 9.5.3.3.1** 3.1.3 4.3.2.3
Database modeling	**7.4.5.6**
Database recovery	**7.4.5.8**
Datapaths	**9.4.6.2**
De Colmar calculator	9.3.1.2.3
Deblurring (IP)	**7.5.3.4**
Debugging (c)	**4.1.2.5.1 4.2.2.4** 7.2.1.3
DEC PDP series	9.3.3.3.1.3 9.3.3.3.1.5
Decidable propositions	**6.3.6.2 7.1.2.2.4**
Decision properties of grammars	**6.2.3.5**
Decision table	4.1.1.3.6
Decision theoretic model (IR)	**7.3.5.3**
Decision tree	7.2.3.1 7.6.4.1.3
Decoder circuit	**1.5.1.4.9**
DECUS user group	**9.6.2.2.5**
Dedicated applications	**2.2.3.2**
Deduction	**7.2.2**
Definite languages	6.2.2.3
Definite symbolic integration	**7.1.2.5.2**
Degree of a graph	5.2.2.1.6
Deletion from a file	**6.6.2.2** 7.3.2.1.2 7.4
Delimiter (c)	
Demodulation	**3.4.1.2**
Denotational semantics	**6.3.2.2**

Density estimation procedures (PR)	7.6.2.1.5
Deque ('deck')	3.1.2.3.2.2
Design automation	8.2.3.1
Determinant	5.2.4.6 5.2.4 7.2.1.4
Deterministic pattern recognition model	7.6.1.1.1
Developmental (grammar) systems	6.2.7.1
Deterministic simulation	7.7.1.4
Device independence	7.9.6.1
Diagnostic software tools	4.2.2.5
Dictionaries (IR)	7.3.1.3
Difference encoding	3.3.3.3
Difference engine	9.3.1.3.1
Differential analyzer	9.3.2.2.1
Differential equation	5.1.3 5.1.3.7 5.1.3.7.4 7.1.2.2.5
Differentiation, numeric	5.1.2.1
Differentiation, symbolic	7.1.2.3
Differentiator circuit	1.5.2.3
Digit analysis hashing	3.2.3.1.3.5
Digital circuitry	1.5.1
Digital clock	1.1.3.2
Digital computer	1.1.1
Digital computer subsystem	1.2
Digital sorting	7.8.1.1.6.1
Digital-to-analog (D to A) converter	1.5.3.2
Digitization of an image	7.5.1
Digraph	See directed graph
Diminishing increment sort (Shellsort)	7.8.1.1.2.2
Direct access merge	7.8.1.2.1
Direct access to a DBMS	7.4.3.3
Direct coupled modem	1.4.5.1
Direct file	3.2.3.2.2.3
Direct solution of linear equations	5.2.4.1
Directed graph (digraph)	5.2.2.1.1 3.1.2.6.1 7.2.6.1
Discourse analysis	7.2.4.3
Discrete event simulation models	7.7.2
Discrete mathematics	5.2 9.2.1.2.4
Disk cartridge	1.4.3.2.2
Disk pack	1.4.3.2.3
Display devices	1.4.3.4 7.9.1.1 8.4.2.3.3
Dispatching	9.5.5.3.3
Distance measures (PR)	7.6.2.2.1
Distinguished vertex (root)	5.2.2.4
Distortion	3.4.1.4.1.2
Distributed processing	1.3.2.2.3 2.1.2.1 9.5.3.4.2
Distribution of resources	8.1.4
Distribution sort	7.8.1.1.6
Distributive external sort	7.8.1.2.3
Dithering (CG)	7.9.7.5
Divided differences	5.1.6.1.3 5.1.6.1.2
Division key hashing	3.2.3.1.3.1
Doctoral programs in computer science	9.2.1.1
Document abstracts (IR)	7.3.1.2
Document display (IR)	7.3.3.4.2

INDEX

Documentation	**4.1.3.4 9.5.1.3.4**
	9.5.2.2.5 9.5.5.4.4 4.1.2.5.1
Domain transformation	**8.2.2.1.3**
Domination, graph	**5.2.2.1.4**
Doubly linked list	**3.2.3.1.2.2**
Do-until loop	See structured programming (u)
Do-while loop	See structured programming (u)
DPMA	**9.6.2.1.3**
Drum, magnetic	**1.4.3.2.4**
Dual graphs	**5.2.2.3.3**
Dynamic redundancy	**1.7.3.2.2**
Dynamic restructuring of a DBMS	**7.4.5.4**
Dynamic scoping	**6.3.3.1.2.3**
Dynamo simulation language	**7.7.3.3.1.3**
EBCDIC character code	**3.2.1.2.1**
Eccles-Jordan circuit	**9.3.2.3.1**
Echoing	**3.4.1.4.1.4**
Eckert, J. Presper	**9.3.2.3.4**
Edge of a graph	**5.2.2**
Edge detection (IP)	**7.5.6.2**
Edsac	**9.3.2.3.5**
Education	**9.1.3.5.2 9.2**
	9.5.4.2.3 7.1.4.6
Educational uses of computers	**9.2.2**
Edvac	**9.3.2.3.5**
Effectiveness of software	**4.1.3.5 8.3.5.2**
Efficiency of algorithms	**5.3.1.2**
Egoless programming	See software engineering (u)
Eigenvalue	**5.2.4.5** 5.2.4
Electromechanical computers	**9.3.2.2**
Electronic funds transfer (EFT)	**2.3.4.6.3 9.4.8.3**
Electronic mail	**8.4.3.1**
Electrostatic printer	**1.4.2.1.3**
Element improvement	**1.7.3.1**
Elementary function	**5.1.6.2.7**
Elliptic PDE	**5.1.4.1**
Embedded data manipulation language (DBMS)	**7.4.4.5.2**
Employment statistics	**8.1.2.3 9.1.1.1 9.7.4.2**
Emulation (c)	
Encryption of data	**3.3.4**
Engineering data processing	**8.2**
Engineering, algebraic manipulation in	**7.1.4.5**
Enhancement of an image	**7.5.3**
Enhancement of software	**4.1.3.6.2 9.5.2.4.2**
Eniac	**9.3.2.3.4**
Ensemble processor	**1.3.2.1.2**
Entity	**7.7.2.2.3** 3.2.3
Enumeration sort	**7.8.1.1.4**
Enumeration, tree	**5.2.2.4.3**
Equalization (conditioning)	**3.4.1.4.2**
Ergonomics	**7.9.6.5**
Error analysis	**5.2.4.2 5.3.1.3 8.2.2.2**

Error detection/correction	4.1.1.4.4 4.2.3.2.2
	9.5.2.4.1 3.4.1.4.3.1-2
Error handling in software	4.2.1.4.3 4.2.3.1.2
Ethics	9.6.3 9.7.7.1.1
Euclid's algorithm	7.1.2.1.2
Euler-Maclaurin formulas	5.1.2.3
Eulerian path and circuit of a graph	5.2.2.2.2
Euler's formula	5.2.2.3.1
Evaluation of new systems	9.5.3.1.1
Event-oriented simulation languages	7.7.2.3.1.2
Excess value exponential notation	3.2.1.1.2.1
Exchange sort	7.8.1.1.3
Execution of instructions	1.2.1 1.2.1.2
	1.2.2.1.2 4.1.1.4.1.2
Exec-8 operating system	9.3.3.3.2.1
Existential instantiation (AI)	7.2.2.1
Exogenous/endogenous simulation variables	7.7.2.2.5.3
Experimental data analysis	8.2.2
Exploration robots	7.2.7.1
Exploratory data analysis (PR)	7.6.2.1.4
Exponent	5.2.1.2 3.2.1.1.2
Exponential space complexity	6.5.2.3
Exponential time complexity	6.5.1.4
Expression (c)	7.1.1
Expressiveness of programs	6.3.6.1
Extensible language	4.1.1.3.5
External file	3.2.3.2
External sorting	7.8.1.2
Extrapolation	5.1.3.3
Facilities management	9.4.5.4.2
Fact retrieval systems (IR)	7.3.6.1.2
Factor analysis	8.2.2.3.4
Factorization of matrices	5.2.4.1.3
Factorization of polynomials	7.1.2.1.3
Fast Fourier Transform (FFT)	5.1.6.2.3 6.6.1.1
Fast Poisson solvers	5.1.4.1.5
Fault classification (hardware)	1.7.2
Fault simulation	1.7.5
Fault-tolerant systems	2.3.2.2
FCC inquiries	9.4.6.1
Feasibility studies	9.5.2.1.3
Feature extraction/evaluation	7.6.2.2 7.6.4.1.2
Feature measurement (IP)	7.5.7
Fetch, memory	1.2.2.2.1.2
Feynman diagram	7.1.4.1.2
Fibonacci search/storage	7.8.2.2.2 3.3.1.5.2
Field (c)	4.1.1.3.8
	4.3.1.2.3 4.3.2.1 7.8
Field-effect semiconductor element	1.6.1.2
FIFO queue	3.1.2.3.2.1
File	3.2.3 4.1.1.5.2.4
	7.8.3.2 4.1.1.3.8
File organization and management	4.3.1 3.2.3 7.4.5.2 7.3

INDEX

Filtering (IP)	7.5.3.4.1
Financial systems and data	8.1.1 9.5.1.2
	9.1.1.3 9.5.1.1.4 9.5.1.2
Finite differences	5.1.2.1 5.1.3.7.2
	5.1.4.1.1 5.1.4.2.1
	5.1.4.1.3 7.1.4.3.3.2
Finite element method	5.1.3.7.5 5.1.4.1.2
	5.1.4.2.5 7.1.4.3.3.2
Finite-state machine or automata	6.1.4 6.1.3 6.4.3 6.1.2
	7.6.1.2.2 7.9.6.4.4 6.2.2.1
Firmware	2.4.2.1.2.3 7.9.1.1.7 7.1.3.7
	See also emulation (c)
First fit storage management	3.3.1.2
First generation computers	9.3.3.1
First normal form (DBMS)	7.4.1.2
Fitch's formal logic system	7.2.2.1
Fixed length record	3.2.2.1 3.3.2.1
Fixed point iteration	5.1.1.2
Fixed point number	3.2.1.1.1
Flag (c)	
Flat film memory	1.6.2.2.2
Flip-flop	1.5.1.4.1 9.3.2.3.1
Floating point arithmetic	3.2.1.1.2 5.2.1.2
	7.1.1.4.4 3.1.1.2
Floppy disk	1.4.3.2.1
Flow analysis	6.3.5.1
Flow chart (c)	6.3.4.1
Fluid scoping	6.3.3.1.2.3
Folded key hashing	3.2.3.1.3.3
Font generator (CG)	7.9.1.1.7.2
Forensic sciences (IP)	7.5.9.7
Fork/join control structures	4.1.1.4.1.6
Formal algebraic structure	3.1.1.7
Formal languages	6.2
Formal parameter	6.3.3.1.2.1
Formal power series	6.2.3.6
Format/typesetting codes	8.4.2.2.4
Fortran	4.1.1.3.4 7.1.3.3.1 9.3.3.1.4.3
	See also standards (u)
Four-color theorem	5.2.2.3.6
Fourier analysis	5.1.6.2.2
Fourier transform	7.5.2.3 8.2.2.1.3
Fractional part of a real number	3.1.1.2
Frame (CG)	7.9.7.6.3
Free tree	5.2.2.4.2
Frequency distributions	5.2.6.1.3
Frequency division multiplexing	3.4.1.3.2
Frequency domain filtering (IP)	7.5.4.1
Frequency ordered key search	7.8.2.3.2
Front-end processor	1.4.4.1
Functional specifications	9.5.2.1.2
Function, mathematical	7.1.1.2.3-4 5.1.1
Fuzzy set logic	7.6.1.1.3 7.2.2.4

Galerkin's method	5.1.4.1.6 5.1.5.5
Game playing	7.2.5.1
Garbage collection	3.3.1.6
Gasp simulation language	7.7.3.3.1.2
Gaussian elimination	5.2.4.1.1 5.2.4.6.2
Gaussian quadrature	5.1.2.5 5.1.5.6
Gauss-Jordan reduction	5.1.4.1.2
Gauss-Seidel method	5.1.4.1.3 5.2.4.4.2
General purpose computer (u)	
General purpose controller	8.2.1.2.2
General purpose simulation language	7.7.3.3.1.1
General relativity	7.1.4.1.1
Generalized inverse	5.2.4.1.5
Generalized secant method	5.1.1.6.2
Generating function	5.2.3.3
Generating system (grammar)	6.2.1
Gentzen's formal logic system	7.2.2.1
Genus of a graph	5.2.2.3.4
Geometric correction (IP)	7.5.3.2
Givens' method	5.2.4.5.3
Global convergence	5.1.1.8
Global variable	6.3.3.1.1
Global/local simulation variables	7.7.2.2.5.5
Goto	4.1.1.4.1.4
GPSS simulation language	7.7.2.3.1.1
Gradient	5.1.1.6.3 5.1.7.2.3
Graduate education	9.2.1.1
Grammars	6.2.1
Grammatical inference	7.2.3.2
Graph	3.1.2.6 3.1.3.1
Graph theory	5.2.2
Graphic arts/typography	8.4.2.1
Graphic terminal	1.4.2.3 7.9.1.1 8.3.1.3.2
Graphics processing system	2.3.1 7.9 8.2.2.1.2
Gray level (IP)	7.5.3.11 7.5.1.1 8.4.3.2
Greatest common divisor (GCD)	7.1.2.1.2
Greibach normal form	6.2.3.3 6.2.4.1
Group theory	7.1.4.2.1
GUIDE user group	9.6.2.2.2
Halting problem	6.6.5.1 6.3.6.2
Hamiltonian path and circuit of a graph	5.2.2.2.3 6.6.3.2.2
Handshaking	3.4.1.2.2.1
Hard copy	7.9
Hardware	1. 4.1.3.1.2
	7.6.4.1 7.8.4 7.9.1 8.3.1
	9.1.3.1 9.5.3.1
Hardware design studies	2.4.2.3
Hardware improvement studies	2.4.2.2
Hardware measurement tools	2.4.2.1.2.1
Hardware reliability	1.7
Hardwired control unit	1.2.1.2.1
Harvard relay computers	9.3.2.2.4.5
Hashing	3.2.3.1.3 7.8.3.4 7.3.2.1.4

Heap	**3.1.2.5.1.1**
Heapsort	**7.8.1.1.1.3**
Height-balanced tree	**3.1.2.5.1.2**
Hermite polynomial	**5.1.6.1.4**
Hessenberg form of a matrix	5.2.4.5.3
Heterogeneous parallel processor	**1.3.2.2**
Heuristic search	**7.2.5.2**
Hewlett-Packard microprocessor	9.3.3.3.1.6
Hexadecimal (c)	5.2.1.5
Hidden line problem (CG)	**7.9.1.1.7.5 7.9.7.1**
Hierarchical computer system	**2.1.2.2 2.3.3.2**
Hierarchical control structures	**7.2.10.1.1**
Hierarchical data structures	**3.1.2.7**
Hierarchical DBMS	**7.4.2.1.1 3.1.3.3**
Higher educational uses of computers	**9.2.2.1**
High-energy physics (IP)	7.5.9.3
High level language	**4.1.1.3.4** See also operator (c)
Hilbert's formal logic system	7.2.2.1
Histogram	**7.6.2.1.1**
History of computing	**9.3**
Hollerith card (c)	1.4.1.1 9.3.2.1.1.1
Homogeneous array	**3.1.2.2.1**
Homogeneous parallel processor	**1.3.2.1**
Honest complexity classes	**6.5.4.3**
Honeywell 6000 series	9.3.3.3.1.3
Horner's rule	**6.6.1.2.1**
Host computer	1.4.4 7.4.5.10.1 7.9.2.5
Host-embedded data manipulation language	**7.4.4.2.2**
Householder's method	5.2.4.5.3
Huffman tree	**5.2.2.4.5 3.3.2.2.1**
Human factors	**7.4.6.5 7.9.6.5 4.2.3** See also systems analysis (u)
Humanistic studies	9.7.7
Hybrid circuitry	1.5.3
Hybrid computer	1.1.3 2.1.1
Hybrid encryption methods	3.3.4.4
Hybrid redundancy	1.7.3.2.3
Hybrid simulation language	**7.7.3.3.1.2**
Hydraulic storage element	1.6.7
Hydrodynamics, algebraic manipulation in	7.1.4.2.2
Hyperbolic equation	5.1.4.2.3
Hypergeometric probability distribution	5.2.6.3
Hyphenation	**8.4.2.2.1**
IBM card	See Hollerith card
IBM SSEC computer	9.3.2.2.6
IBM 360 & 370	**9.3.3.3.1.2** See also cell (c)
IBM 650	9.3.3.1.2
IBM 700 series	9.3.3.1.1 9.3.3.1.3
IBM 7090	9.3.3.2.2
IBSYS operating system	9.3.3.1.4.3
ICCP	**9.6.2.4**
Identifier	See expression (c) and variable (c)
Identities	7.1.2.2.5

IEEE	9.6.2.1.4
IFIP	9.6.2.1.6
If-then-else structure	See structured programming (u)
Image processing (IP)	2.3.1 7.5 7.6.3.3 7.9.3.2.6
Image storage display device	7.9.1.1.2
Impact device	1.4.2.1.1
Import/export regulations	9.4.8.1
Indefinite integration	7.1.2.5.1
Independent/dependent simulation variables	7.7.2.2.5.4 7.7.3.2.4.4
Index register	See register (c)
Index sequential file	3.2.3.2.2.2
Indexed access array	3.1.2.2.1
Indexed grammar	6.2.5.1
Indexing methods (IR)	7.3.1.1
Indexing of files	4.3.1.2.5 7.4.3.2
Indirect addressing	6.4.2.3.1
Inductive inference (AI)	7.2.3.3
Industrial automation (IP)	7.5.9.4
Inequalities	8.2.4.1.2
Inference methods	7.2.2
Infinite precision integers	7.1.1.4.1
Infinite state machines	6.1.5 See also Turing machines
Information analysis (IR)	7.3.1
Information lossless machines	6.1.2.4
Information science (u)	
Information storage and retrieval (IR)	7.3 7.2.9.4
Information systems (IR)	7.3.6
Information systems education	9.2.1.1.2 9.2.1.2.2
Information theoretic models (IR)	7.3.5.4
Inherent ambiguity in languages	6.2.3.2
Inherently hard problems	6.6.4
Initial value problem	5.1.3.7.7 5.1.4.2
Input/output (I/O)	1.2.3 4.1.1.4.3 4.2.1.3.1 6.4.3.1.1-2 7.1.3.5 7.8.4.3 6.1.2.1
Insertion into a file	6.6.2.2 7.3.2.1.2 7.4
Insertion sort	7.8.1.1.2
Installation of systems	2.4.2.1
Installation standards	9.5.1.3
Installation statistics re computer industry	9.1.1.2
Instruction (c)	1.2.1.2 1.2.1.2.2 1.2.2.1.2 1.4 4.1.1 4.1.1.3.2
Instruction set	6.4.2.3.3
Integer	3.1.1.1 7.1.1.4.1
Integer programming	5.2.5.3
Integral equation	5.1.5
Integral part of a real number	3.1.1.2
Integrated circuit	9.3.3.3.1.1 1.1.1.1 1.5.1.4
Integration, numeric	5.1.2
Integration, symbolic	7.1.2.5
Integrator circuit	1.5.2.2 1.1.1.1 1.5.1.4
Integrity of a system	7.4.2.5 7.4.5.7 7.4
Integro-differential equation	5.1.5
Intel microprocessor	9.3.3.3.1.6

INDEX

Intelligent terminal	1.4.2.4 9.5.3.1.3
Interactive computing	2.2.3 4.2.1.2.2 7.1.3.2.1
	7.3.3.4 7.6.4.2.2 8.4.1.3
	8.4.2.3.2 9.5.3.4.3
	1.4.2 2.2.5 7.9.1.1
Interface	7.4.4 8.3.1.3 3.4.1.2.2.2
	9.5.1.3.3 9.5.3.1.4 1.1.3.3
	1.2.2.1.4 7.4.3 7.9.4
Interframe coding (IP)	7.5.2.5
Interleaved command sequences	6.3.3.2.2.1
Intermediate language	4.1.1.2.2.2
Internal sorting	7.8.1.1
Interpolation	5.1.5.2 5.1.6.1
Interpretation of instructions	1.2.1
Interpreter	4.1.1.2.1.2 4.1.4.4
Interrupt	1.2.3.2.2 7.9.6.4.4
Intersection	6.2.3.4 5.2.2.1.3
Interval arithmetic	5.2.1.4
Invariant embedding	5.1.3.7.7
Inventory control	8.1.5.3
Inverse filtering (IP)	7.5.4.1
Inverse of a matrix	5.2.4.1.5
Inverted file	7.3.2.1.3 3.2.3.2.3.2 3.2.3.1.4
Iteration (c)	5.2.4.3-4 6.3.3.2.1.2 5.1.1.1.2
Iterative array	1.5.1.3 6.1.5.5 6.4.3
I/O addressing	1.2.3.3
I/O channel	See I/O processor
I/O concentrator	1.4.4.2 2.3.3.1.2
I/O control systems	2.3.3.1
I/O device	1.4
I/O path	1.2.3.3
I/O processor (channel)	1.2.3.1.2.2 1.4.4 2.3.3.1
Jacobi's iterative method	5.2.4.4.1 5.2.4.5.2
Job (c)	2.2.1-2
Job control function	4.2.1.1
Journaling	8.4.3.5
Joystick (CG)	7.9.1.2.1 8.3.1.3.3 8.4.1.3.2
Junior college computer science education	9.2.1.3
Justification of text	8.4.2.2.1
Kalman filtering (IP)	7.5.4.4
Karp reducibility	6.6.3.1.2
Kernel graphics system	7.9.6.3
Key	4.3.1.2.4 3.2.3.2
	7.8 3.2.3.1.3
Keyboard/printer terminal	1.4.2.1 8.3.1.3.1 7.1.3.8
Keyword-based query language (DBMS)	7.4.4.3.1
Key-to-disk terminal	8.4.4
Kleene closure	6.2.2.2 6.2.7.2
Knapsack problem	6.6.3.2.4
Knowledge representation (AI)	7.2.6
Kuratowski's graph theorem	5.2.2.3.2
KWIC index (IR)	7.3.1.1.8 7.3.1.3.4

Label (c)	4.1.1.5.1.6
Labor relations	9.4.8.4.1
Lagrangian	5.1.6.1.1
LALR grammar	6.2.4.1
Lambda calculus expressions	6.3.2.1.3 7.1.3.4.3
Lanczos' algorithm	5.2.4.5.5
Language interfaces to a DBMS	7.4.4
Laplace transform	7.5.3.4.2 8.2.2.1.3
Large scale computer	1.1.1.4
Large scale integration (LSI)	9.3.3.3.1.4
Large screen display	7.9.1.1.3
Laser printer/scanner	1.4.1.3.3 7.6.4.1.1 7.9.1.1.5
Latency (c)	7.8.4.3
Lattice semantics	6.3.2.2.4 6.4.3
Laurent series	7.1.1.2.5
Layout of text	8.4.2.3
Leaf of a tree	3.1.2.5.1.1
Learning (AI)	7.2.3
Leasing	9.1.3.6 9.4.5.1.2 9.4.5.2.2
Least squares	5.1.6.2.1 5.1.7.1
	8.2.2.4.1 7.5.4.3
Legal aspects of computing	9.4
Legislative applications	9.7.4.4
Leibniz' calculator	9.3.1.2.2
Lexical scoping	6.3.3.1.2.2
Library of programs	9.5.5.3.4
Licensing	9.4.5.4.3 9.4.8.4.2 9.6.5
LIFO stack	3.1.2.3.1
Light-pen	1.4.2.3 7.9.1.2.2 8.3.1.3.3
Limits, symbolic generation of	7.1.2.4
Lincoln laboratory TZ-0	9.3.3.2.1
Line detection (IP)	7.5.6.4
Line drawing interpretation	7.2.8.1 7.5.9.8 7.6.3.2
Line printer	1.4.1.3.1
Linear algebraic system	5.2.4 7.1.2.1.4 5.1.4.1.3
Linear bounded automata (LBA)	6.2.6.1
Linear congruential random number generator	5.2.6.1.1
Linear language	6.2.5.2
Linear list	3.1.2.3
Linear programming	5.2.5.1 6.6.3.2.4
Linear space complexity	6.5.2.1
Linear time complexity	6.5.1.2
Linguistic processing	7.3.1.4
Linkage editor	4.2.2.1
Linked access to storage	7.4.3.4 3.2.3.1.2
Lisp	7.1.3.3.2
List	3.1.2.3 4.1.1.5.2.2 7.8.2.4
	3.1.2.4.2 3.1.2.6.1.2
	7.1.1.1 7.8.1.1.2.3
Literature, computers in	9.7.7.2
Litigation	9.4.9
Load balancing	8.1.4.2
Loader program	4.2.2.2
Local variable	6.3.3.1.2

INDEX

Location, memory	1.2.2.2.1 See also cell (c)
Logic circuitry	1.5.1
Logic (program)	6.3.4
Logical database	3.1.3
Logical design of a DBMS	7.4.2
Logical (Boolean) data type or expression	3.1.1.3 3.2.1.4
	4.1.1.5.1.2 6.6.3.2.1
	See also operator (c)
Logical storage allocation	4.3.1.2
Long range planning of computer resources	9.5.1.1
Loop (c)	4.1.1.4.1.2 6.3.3.2.1
Loosely-coupled computer system	2.1.2.2.2
LR grammars	6.2.4.1
Machine intelligence	9.7.4.5 7.2
Machine language (c)	4.1.1.2
Machine readable form	9.4.2.1-2
Macro grammar	6.2.7.1
Macro instruction	4.1.1.3.3
MACSYMA	7.1.3.3.2
Magnetic bubble memory	1.6.2.3
Magnetic card	1.4.3.3
Magnetic core	1.6.2.1 9.3.3.1.3
Magnetic disk	1.4.3.2.1-3 1.2.2.1.3 4.3.1.1
Magnetic drum	1.4.3.2.4 1.2.2.1.3
Magnetic flat film	1.6.2.2.2
Magnetic plated wire	1.6.2.2.1
Magnetic storage elements	1.6.2
Magnetic tape	1.4.3.1 1.2.2.1.3
	4.3.1.1 8.2.1.1.1-2
Magnetic thin film	1.6.2.2
Main memory	1.2.2.1.2 4.2.2.2
Maintenance	4.1.3.6 9.1.3.5.6
	9.4.5.3 9.5.2.4 9.5.5.4
Management information system (MIS)	8.1.6
Management of computing	9.5
Manipulation robots	7.2.7.2
Mantissa	7.1.1.4.5 3.2.1.1.2
Manufacturing systems	8.1.5
Manuscript editors	8.4.1.2
Man-machine interaction	7.9
Mapping between levels of DBMS	7.4.4.1.4
Mark I computer	9.3.2.2.4
Mark II computer	9.3.2.2.5
Marketing systems and data	8.1.3 9.1.2
Markov algorithm	7.1.3.4.2 7.1.3.3.3
Mark-up systems	8.4.2.2
Mass memory storage device	1.2.2.1.3 1.1.1.2 1.1.1.4
Masters programs in computer science	9.2.1.1
Master-slave computer system	2.1.2.4
Mathematical programming	5.2.5
Mathematical statistics	5.2.6
Mathematics, algebraic manipulation in	7.1.4.2
Mathematics of computing	5.

Matrix	3.1.2.2
Matrix multiplication	**6.6.1.4**
Matrix printer	**1.4.2.1.2**
Mauchly, John	9.3.2.3.4
Mealy circuit model	**1.5.1.2.1**
Mean-time-between-failures	1.7.1
Mean-time-to-failure	1.7.1
Measurement techniques	**2.4.2.1.2**
Mechanical storage element	**1.6.6**
Median	8.2.2.4.5
Medicine, computer applications to	7.1.4.7 **7.5.9.2**
Medium scale computer	**1.1.1.3 9.3.3.1.2**
Memory	**1.2.2**
Memory access	**1.2.2.2** 1.2.3.1.2.2
Memory based I/O	**1.2.3.1.1**
Memory constrained algorithms	**5.3.1.4**
Memory mapped I/O	**1.2.3.3.2**
Menu type query language (DBMS)	**7.4.4.3.5 7.9.6.4.1**
Merge	**7.8.1.1.5 7.8.1.2.2** 5.2.2.4.5
Merge exchange sort	**7.8.1.1.3.2**
Mesh	5.1.2.5 5.1.4.1.2
Message switching	**1.4.4.3 2.3.4.4 3.4.2.2.1**
Metalinear language	6.2.5.2
Method of characteristics	**5.1.4.2.3**
Method of lines	**5.1.4.2.2**
MICR	See standards (u)
Microcode	2.4.2.1.2.3
Microcomputer	**1.1.1.1**
Microfilm recorder	**1.4.1.6 7.9.1.1.5** 9.1.3.1.6
Microinstruction	1.2.1.2.2
Microprocessor	**9.3.3.3.1.6 9.5.3.1.2**
Microprogram	1.2.1.2.2
Microprogrammed control unit	**1.2.1.2.2**
Microsecond (c)	
Microwave data transmission	3.4.1.1
Midsquare hashing	**3.2.3.1.3.2**
Millisecond (c)	
Minicomputer	1.1.1.2 7.9.2.4
	9.3.3.3.1.5 9.5.3.1.2
Minimax polynomial	**5.1.6.2.4**
Minimization of nonlinear functions	**5.1.7.2**
Minimum spanning tree	**5.2.2.4.4**
Minor sequences in computer science education	9.2.1.2.5
Mobius inversion	5.2.3.4.2
Modal logic	**6.3.4.5** 7.2.2.4
Model validation	7.7.4.1
Modeling techniques	7.7 2.4.2.1.1 4.1.3.3.1
Modem	1.4.5 3.4.1.2
	3.4.1.2.2 9.1.3.1.5
Modem interface control circuits	**3.4.1.2.2.2**
Modulation	3.4.1.2
Module	**4.1.3.3.2 9.5.2.2.3**
	See also software engineering (u)
Modus ponens	7.2.2.1

Monoalphabetic substitution	3.3.4.2
Monte Carlo method	**5.2.6.2 8.2.4.1.3**
Moore circuit model	**1.5.1.2.2** 1.5.1.4.1
Motion picture analysis (AI)	**7.2.8.3**
Motorola microprocessor	9.3.3.3.1.6
Muller's method	**5.1.1.5**
Multiaperture magnetic core	**1.6.2.1.2**
Multics time sharing system	9.3.3.3.2.2
Multidimensional memory access	1.2.2.2.1.3
Multidimensional tape Turing machine	**6.4.1.2.3**
Multihead tape Turing machine	**6.4.1.2.2**
Multilinked storage structure	3.2.3.1.2.3
Multilist file	7.3.2.1.2 3.2.3.2.3.1
Multiple correlation	**8.2.2.3.2**
Multiple fault	**1.7.2.1.2**
Multiple precision arithmetic	3.2.1.1.2.2
Multiple processing units	**2.1.3.3** 1.3.2
Multiplexing	**3.4.1.3 1.5.1.4.8**
	See also channel (c)
Multiplier circuit	1.5.1.4.6
Multiprocessing	**1.3.2.1.4** 2.1.2
	4.2.1.3.5 7.8.4.4.2
Multiprogramming	**4.2.1.3.4** 9.3.3.3
Multistep query refinement (IR)	**7.3.3.4.3**
Multistep soln of ord diff equations	**5.1.3.2**
Multitape automata	**6.1.4.2 6.4.1.2**
Nanosecond (c)	
Napier's bones	9.3.1.1.3
Natural language processing	**7.2.4 7.3.3.1.2** 7.2.1.3
NCR 304	9.3.3.2.1
Nesting of loops	6.4.2.2
Network	1.3.2.2.4 2.3.4.5 5.2.2.4.8
	7.2.3.1 **7.2.6.1** 7.9.2.5
	9.5.1.1.5 9.5.3.2.3 9.5.3.4.1
Network DBMS	7.4.2.1.2 3.1.3.1
Neural nets	**6.1.2.5**
Newton's method	**5.1.1.4** 5.1.7.2.1
Newton-Cotes formulas	**5.1.2.2.3**
Next fit storage management	3.3.1.4
Node of a graph or tree	5.2.2 3.1.2.5
Noise	3.4.1.4.1.1
Noise cleaning (IP)	**7.5.3.3.1**
Noncounting language	6.2.2.3
Nondeterministic machines	6.1.3.2 **6.2.3.1**
Nondeterministic polynomial (NP) time	6.6.3.2
Nondeterministic simulation	7.7.1.4
Nonexecutable program statements	4.1.1.4.5
Nonimpact printing devices	**1.4.2.1.3**
Nonlinear list	3.1.2.4
Nonlinear programming	**5.2.5.2**
Nonnumeric programming (u)	
Nonprocedural language	4.1.1.3.7 7.1.3.3.3
Nonstandard logics	7.2.2.4

Normal distribution	8.2.2.2
Normalization (DBMS)	**7.4.1.2**
NP-complete problems	**6.6.3.2**
Number systems	5.2.1
Numeric storage structures	3.2.1.1
Numerical analysis, algebraic manipulation in	7.1.4.3.3
Numerical differentiation	**5.1.2**
Numerical integration	**5.1.2**
Numerical software	**5.3.2**
N-ary tree	3.1.2.5.2 7.8.2.2.3
Object program or code (c)	4.1.1.2.2.4
Occupational titles	**9.6.1**
Octal (c)	5.2.1.5
Odhner calculator	**9.3.1.4.2**
Office automation	7.3.6.2.4 8.4.3 **9.5.3.3.3**
Off-line	**6.4.1.3**
One-sided linear (type 3) grammars	**6.2.1.4** 6.2.2
One-tape Turing machine	**6.4.1.1**
Ones complement number system	3.2.1.1.1.3
On-line	**6.4.1.3 8.4.1.4** 1.4.2
Open subroutine	See subroutine (c)
Operand (c)	
Operating system	**4.2.1 9.3.3.3.2.1**
	9.6.3.2.1 4.1.4.3
	4.2.1.3.5 7.4.5.5 9.3.3.3
Operation code (op code) (c)	
Operational amplifier	**1.5.2.1**
Operational semantics	**6.3.2.1**
Operator (functional) (c)	
Operator (human)	4.1.3.4.2.3 4.2.3.1
Optical character recognition (OCR)	**7.6.3.2.1** 8.4.4
	See also standards (u)
Optical scanner	**1.4.1.5**
Optical storage element	**1.6.5**
Optimization	**5.1.7** 1.1.3.1 **6.3.5**
Optimization of object code	**4.1.1.2.2.4**
Oracle	6.6.5.3
Order entry	**8.1.3.3**
Ordinary differential equation (ODE)	**5.1.3** 7.1.2.5.3
Organization (architecture) of a computer	1.3
Orthogonal function	**8.2.2.4.4** 5.1.4.1.6
	5.1.5.3-4 5.2.4.5.5-6
Orthonormal expansion	**5.1.5.3**
OS 360 operating system	9.3.3.3.2.1
Overlap of I/O and computing	**7.8.4.3.2**
Overlay	4.1.1.4.2
Overrelaxation method	**5.1.4.1.3** 5.2.4.4.5
Packet switching	**9.1.3.4.1 3.4.2.2.2**
Page fault	6.6.1.4.2
Pagination of text	**8.4.2.3**
Pan's matrix multiplication algorithm	6.6.1.4.1
Paper tape reader/punch	**1.4.1.2** 8.2.1.1.2

INDEX

Parallel language control structures	4.1.1.4.1.6 6.3.3.2.2
Parallel processing	1.3.2 2.1.3 4.2.1.3.3 7.8.4.4.1
Parallel rewriting systems	6.2.7.1
Parameter	1.1.3.1 4.1.2.1
	4.2.2.1 6.3.3.1.2.1
	7.2.3.4 8.2.2.4.5
Parity	4.3.1.1.3 3.4.1.4.3
Parsing	6.2.4 7.2.4.2
	7.1.3.5.3 7.2.3.2
Partial correctness logic	6.3.4.2
Partial differential equation (PDE)	5.1.4 7.1.4.3.3.2
	5.1.4.1.6 7.1.4.1.3
Partial fraction expansion	7.1.2.1.6
Partially ordered set (poset)	5.2.2.2.6
Partitioned database	1.3.2.2.3
Partition exchange sort (quicksort)	7.8.1.1.3.1
Pascal's calculator	9.3.1.2.1
Passive instrumentation	8.2.1.1
Patchable hybrid computer	1.1.3.1-2
Patents	9.4.1.2
Path in a graph	5.2.2.2
Pattern analysis	7.6.2.1
Pattern classification experiments	7.6.2.4
Pattern matching	7.1.2.2.3 7.5.6.3
	8.4.1.1.2 3.1.1.7
Pattern recognition	6.4.3.3 7.6 7.9.6.4.5
Pattern substitution encoding	3.3.3.2
Payroll applications	8.1.2.2
Performance of systems	2.4 4.1.3.2.2 9.5.5.5.1
Peripheral devices and subsystems	1.4.3 2.3.3.1.1 9.1.3.1.2
	9.5.5.2.2 1.1.1.4
	4.2.3.1.3 9.5.3.1.4
Permanent fault	1.7.2.2.1
Permutations	5.2.3.1.1
Personal computer	8.3.1.1
Personnel systems	8.1.2 9.5.1.1.3 9.5.6.4
Perturbation techniques	7.1.4.2.2
Petri net	6.3.2.1.5 6.3.3.2.2.2
Phase error	3.4.1.4.1.2
Philco Transac S-2000	9.3.3.2.2
Philco 1000	9.3.3.2.1
Phrase dictionary (IR)	7.3.1.3.3
Phrase formation (IR)	7.3.1.1.5
Phrase structure grammar	7.3.1.4.1
Physical database	3.2.4
Physical storage allocation	4.2.1.3.2.1 4.3.1.1
Physics, algebraic manipulation in	7.1.4.1
Picture generation (CG)	7.9.3
Picture processing	7.5
Piecewise linear hashing	3.2.3.1.3.6
Piecewise polynomial interpolation	5.1.5.2
Pipeline machine	1.3.2.2.1 2.1.3.1 6.6.1.6
Pixel (IP)	7.5.1 7.5.2.2 7.9.1.1.7.6
Placement of personnel	9.1.3.5.5

Planar graph	**5.2.2.3** 5.2.2.3.4
Planner (AI)	7.2.10.2
Plaintext	3.3.4.2
Plasma display device	**1.4.3.4.3**
Plasma physics	**7.1.4.1.3**
Plated wire memory	**1.6.2.2.1**
Plotter	**1.4.1.4 7.9.1.1.4 7.9.4.3.1**
PL/1	7.4.4.2.2 6.3.2.1.2
	See also standards (u)
Pointer (data type)	**3.1.1.6 3.2.1.3 4.1.1.5.1.5**
	7.3.2.1.2 7.4.5.2.2 7.8.2.4
Point-of-sale transaction	**2.3.4.6.2**
Poisson distribution	5.2.6.1.3 8.2.2.2
Political campaign practices	**9.7.4.4.2**
Polling	**3.4.1.3.3.4**
Polyalphabetic substitution	3.3.4.2
Polya's theory	**5.2.3.5**
Polynomial	**7.1.1.2.1** 5.1.1.7 5.1.5.2
	5.1.6.1 5.1.6.1.4 5.1.6.2.1
	5.1.6.2.4 5.1.6.2.8 6.6.1.2
Polynomial space complexity	**6.5.2.2**
Polynomial time complexity	**6.5.1.3 6.6.3.1**
Portability of software	**4.1.4 5.3.2.1 7.7.4.4** 7.9.6.1
Post correspondence problem	**6.6.5.2**
Post's variant of a Turing machine	**6.4.1.1.2**
Power iteration eigenvalue calculation	**5.2.4.5.1**
Power series	**7.1.1.2.5**
Powers' punched card machine	**9.3.2.1.1.2**
Precedence grammars	**6.2.4.1**
Precision	3.2.1.1.2.2 7.1.1.4.1 7.1.1.4.5
Precision (IR)	**7.3.4.2.2**
Precollege education in computer science	**9.2.2.2**
Preconditioning	**6.6.1.2.2**
Predicate calculus	**7.2.6.2** 6.3.6
Predictive coding (IP)	**7.5.2.2**
Predictor-corrector formulas	**5.1.3.2.1**
Prefix codes	5.2.2.4.5
Preprocessor	**4.1.1.2.2.3**
Presburger arithmetic	6.6.4.2
Primary key	**3.2.3.2.2**
Primitive instruction	3.1.1 3.2.1 7.9.1.1.7
Printer	**1.4.1.3**
Priority queue	**3.1.2.3.2.3**
Privacy of system use	**4.2.1.4.1 9.4.2 9.7.4.3** 3.4.3.1
Probabilistic analysis	**7.1.2.7.4** 7.3.5.2
Probabilistic automata	**6.1.4.3**
Probabilistic models (IR)	**7.3.5.2**
Probability	**5.2.6** 9.2.1.2.4
Problem-oriented language	**5.3.2.6**
Problem solving	**7.2.5 8.3.3.6**
Procedural query languages (DBMS)	**7.4.4.3.4**
Procedure (c)	4.1.1.4.1.5 6.3.3.1.2.1
	6.3.3.2.1.3 **4.1.2.1**
Procedure-oriented language (POL)	**7.1.3.3.2**

Processor	1.2.3 4.1.1.2.1
Process control	**2.3.4.2**
Process-oriented simulation languages	**7.7.2.3.1.3**
Procurement of systems	2.4.2.1 **9.4.8.2**
Product profitability analysis	**8.1.6.2**
Production scheduling	**8.1.5.2**
Professional ethics	**9.6.3**
Professional services	**9.5.4.2 9.6 9.1.3.5**
Program (c)	4. **4.1.3 7.2.1.1 7.8.3.3**
Program construct	**6.3.3**
Program editor	**8.4.1.1**
Program libraries	**9.5.5.3.4**
Program logic	**6.3.4**
Program optimization	**6.3.5**
Program semantics	**6.3.2**
Program syntax	**6.3.1**
Program schematology	**6.3.6**
Program verification	See verification of programs
Programmed grammars	**6.2.7.1**
Programmed I/O	**1.2.3.2.1**
Programmer	**9.6.1.4 4.1.3.4.2.2 9.2.1.3.1** 1.2.3.1.1 4.2.2.4
Programmer-defined data type	4.1.1.5.1.6
Programming	**9.5.2.2.2 9.5.3.2.2 9.5.4.2.2**
Programming assistants	**7.2.1.3**
Programming instruction	**9.2.1.4.1**
Programming languages	**4.1.1**
Project management	**9.5.2.5**
Projections of an image	**7.5.7.4 7.9.3.2.3**
Protection of software	**4.2.1.4**
Protection specification languages	**7.4.4.4**
Protocol	**3.4.1.4.4**
Pseudoinverse restoration (IP)	**7.5.4.3**
Pseudoinverse, matrix	**5.2.4.1.5**
Public key cryptosystem	**3.3.4.3**
Public policy issues	**9.7.4**
Pulse-code modulation	**3.4.1.2.1**
Punched card	1.4.1.1 4.2.3.2.2 See also Hollerith card (c)
Purchasing systems	**8.1.7.1**
Pushdown automata	**6.1.5.2**
Quadratic selection sort	**7.8.1.1.1.4**
Quality control	**8.1.5.4 9.5.5.3.5**
Quantization of a transmitted image	**7.5.1.2**
Query formulation (IR)	**7.3.3.1** 7.3
Query languages (DBMS)	**7.4.4.3**
Query-record simulation (IR)	**7.3.3.3.3**
Question answering systems	**7.3.6.1.2** 7.2.2
Questionnaire type language	**4.1.1.3.6**
Question-and-answer sequences (CAI)	**8.3.4.1**
Queue	**3.1.2.3.2**
Quicksort	**7.8.1.1.3.1**
Quotient-difference algorithm	5.1.1.7

Q-R method	5.2.4.5.4
Radix (c)	**5.2.1.5** 3.2.1.1.1.2-3
Radix exchange sort	7.8.1.1.3.3
Radix sort	**7.8.1.1.6.1**
Radix transformation hashing	**3.2.3.1.3.4**
Radon transform (IP)	**7.5.5.2**
Ramsey theory	**5.2.3.7**
Random access memory	**1.2.2.2.1.1**
Random number generation	**5.2.6.1** 7.7.4.2
Raster display device	**7.9.1.1.1.2**
Rational fraction or function	**5.1.6.2.5 7.1.1.2.2 7.1.1.4.2 7.1.2.5.1.1 8.2.2.4.3**
Rayleigh-Ritz method	**5.1.3.7.6**
RCA Bizmac	9.3.3.1.3
RCA 501	9.3.3.2.1
Read-write head	6.4.1.1.1
Real number	**3.1.1.2** 5.1.1 5.2.1.1
Real time	**4.2.1.2.1 6.5.1.1 9.5.3.4.3** 2.3.4.1-2 2.3.4.7 7.9.6.4.5
Recall (IR)	**7.3.4.2.1**
Reconnaissance (IP)	**7.5.9.6**
Reconstruction of an image	**7.5.5**
Record	**3.2.2 4.1.1.5.2.3** 7.8 **4.3.1.2.3 7.3.2.2 7.8.1.2.3.1** 1.2.2.2 4.1.1.3.8 7.4.2.1.1
Record assignment	**6.3.3.1.1.3**
Record-record similarity (IR)	**7.3.3.3.3**
Recruiting	9.1.3.5.5
Recovery (DBMS)	7.4.5.8
Recurrence relations	**5.2.3.2**
Recursion	**4.1.2.3 6.3.3.2.1.4 7.1.1.3.1.2 7.2.10.1.1.1** 6.3.5.2
Recursive filtering (IP)	**7.5.4.4**
Reel tape device	1.4.3.1.3
Reentrant code	4.1.2.4
Reference retrieval systems (IR)	**7.3.6.1.1**
Refresh graphic devices	7.9.1.1.1
Refutation methods (AI)	7.2.2.2
Region growing (IP)	**7.4.6.6**
Register (c)	1.2.2.1.2 1.2.3.1.1
Register based I/O	1.2.3.1.1
Registration of software	9.4.1.4
Regression analysis	8.2.2.3.4
Regular expression	**6.1.3.3** 6.2.2.2 **6.6.4.1**
Relational DBMS	**7.4.2.1.3** 3.1.3.2
Relative storage location	3.2.1.3.2 4.2.2.2
Relay	1.5.1.1.1
Relay computers	9.3.2.2
Reliability	1.7 5.3.2.5 8.3.5.3 9.7.3.6
Remote computing service	9.5.4.1.2 9.5.5.2.3
Remote concentrator	1.4.4.2
Remote job entry	2.2.2
Remote sensing (IP)	7.5.9.5

Replacement statement	See expression (c)
Representation of a data structure	**3.2**
Reservation-oriented computer system	**2.3.4.6.1**
Residue calculation	7.1.2.5.2
Resolution principle	7.2.2.2
Response time (c)	**2.4.1.2**
Restart procedure	**4.1.2.5.2** 7.4.5.8
Restoration of an image	**7.5.4**
Restricted context-free language	6.2.5
Restricted context-sensitive language	**6.2.6.2**
Retrieval evaluation (IR)	**7.3.4**
Retrieval models (IR)	**7.3.5**
Retrieval of information	7.3.3.3 **7.3.4-5**
Retrieval time	1.2.2.2.1.1
Rewriting rules	**6.2.7.1** 6.2.1.1
Robotics	**7.2.7** See also cybernetics (u)
Robustness of software	**5.3.2.5**
Romberg integration	**5.1.2.4**
Root node of a tree	5.2.2.4 3.1.2.5
Root of an equation	5.1.1.7
Rotational delay (latency)	7.8.4.3
Roundoff error (c)	**5.2.1.1** 5.2.1.3-4 5.2.4.2
Run length encoding	**7.5.2.1.2** 3.3.3.1
Runge-Kutta Fehlberg method	**5.1.3.1.3**
Runge-Kutta method	**5.1.3.1.2**
Sail (AI)	7.2.10.2
Sales analysis	**8.1.3.2**
Sales forecasting	**8.1.3.1**
Sampling of an image	**7.5.1.1**
Sap assembler	9.3.3.1.4.1
Satellite, communications	**9.4.6.2.1** 2.3.4.7
Satisfiability of a Boolean expression	**6.6.3.2.1**
Scatter diagram	**7.6.2.1.1**
Scatter storage	**7.3.2.1.4**
Scene analysis (IP)	**7.2.8.2** 7.5.8
Scheduling of programs	**4.2.3.1.1** 9.5.5.3.2
	4.2.1 2.4.2.2
Schema (DBMS)	**7.4.4.1.1** 7.4.1.3
Schickard calculator	9.3.1.2.3
Scientific data processing	**8.2**
Scientific notation	5.2.1.2
Schmitt trigger circuit	**1.5.3.3**
Scope operating system	9.3.3.3.2.1
Scoped assignment	**6.3.3.1.2**
Scratchpad	7.1.3.3.3
Screen copiers	**7.9.1.1.6 8.4.1.3.2**
SDI system (IR)	**7.3.6.1.3**
Searching	**6.6.2.1 7.8.2**
	1.3.2.1.1 5.2.2.4.5
Search tree	**5.2.2.4.6** 7.8.2.2
Secant method	**5.1.1.3**
SECD machine	**6.3.2.1.1**
Second generation computers	**9.3.3.2**

TAXONOMY OF COMPUTER SCIENCE AND ENGINEERING

Secondary school education	9.2.1.4
Security	4.2.1.4.2 7.4.5.9 7.4.6.1 9.4.3
	9.5.5.1 3.4.3.1 7.4
Segmentation of images	7.5.6 7.6.1.2.1
Selection of systems	2.4.2.1
Selection sort	7.8.1.1.1
Selector channel	See channel (c)
Self-assessment	9.6.4.2 9.2.1.5.2
Self-organizing systems	2.3.2.1
Self-reproducing machine	6.4.3.3
Semantic network	7.3.1.4.6
Semantic representation	7.4.2.3
Semantics	4.1.3.7.2 4.1.1.1.2
	4.1.1.3.5 6.3.2
Semiconductor element	1.6.1
Sentence extraction (IR)	7.3.1.2.1
Sentence generation (AI)	7.2.4.4
Sentence scoring (IR)	7.3.1.2.2
Sentinel	See flag (c)
Sequencing	See structured programming (u)
Sequential access	1.2.2.2.1.2
Sequential circuit	1.5.1.2 1.5.1.3
Sequential control	6.3.3.2.1
Sequential file	7.3.2.1.1 3.2.3.1.1
Sequential processor	1.3.1 6.1.2
Sequential search algorithm	7.8.2.1 3.2.3.2.2.1
Serial I/O	6.4.3.1.1
Serializable concurrency	6.3.3.2.2.1
Series expansion of images	7.5.5.3
Service bureaus	9.1.3.3.1
Service courses in computer science	9.2.1.1.4 9.2.1.3.3
Set	3.1.2.1
Set theoretic models	7.3.5.1 7.4.2.1.4
Shannon-Fano-Huffman encoding	7.5.2.1.1 3.3.2.2.2
SHARE users' group	9.6.2.2.1
Shared logic word processing systems	8.4.1.2.2
Shared memory computer systems	2.1.2.3
Sharpening of images	7.5.3.4
Shellsort	7.8.1.1.2.2
Shepherdson-Sturgis machines	6.4.2.1
Shift	6.4.1.1.1-3
Shift register	1.5.1.4.2
Shorted fault	1.7.2.2.1.2
Shortest path problem	5.2.2.2.4
Shooting methods	5.1.3.7.1
Shrinking of images	7.5.6.7
Shuffle operator	6.3.3.2.2.1
Sign-magnitude number system	3.2.1.1.1.1
Signal	1.2.1.2
Signal processing	2.3.4.7 7.1.4.5
Signaling mechanism	1.2.3.2
Significant digit arithmetic	5.2.1.3
Simplex method	5.2.5.1.1
Simplification of an algebraic expression	7.1.2.2

INDEX

Simpson's rule	5.1.2.2.2
Simscript	4.1.1.3.4 7.7.2.3.1.2
Simula	7.7.2.3.1.3
Simulation	7.7 4.1.3.3.1
	8.2.3 2.4.2.1.1.1 8.3.3.4
Simulation languages	7.7.2.3
Single fault	1.7.2.1.1
Single linked list	3.2.3.1.2.1
Single precision arithmetic	3.2.1.1.2.2
Singular value decomposition	5.2.4.5.6
Skills inventory	8.1.2.1
Slide rule	9.3.1.1.2
Smoothing of data	8.2.2.1.1 7.5.3.3
Snobol	3.1.2.2.2 4.1.1.3.4
Soap assembler	9.3.3.1.4.1
Social issues of computing	9.7
Software	4. 4.1.3.1.3 5.3
	5.3.2.3 7.2.10 7.4.5.1
	8.3.2 9.1.3.2 9.4.5.2
	9.4.1 9.5.5.4.2 9.5.1.3.2
	9.5.3.2 9.5.5.4.3
Software engineering (u)	
Software monitor	4.1.3.5.2
Software tools	2.4.2.1.2.2 4.1
Sorting	6.6.2.3 7.8.1
Source data automation	7.3.6.2.1
Source program (c)	4.1.1.2.1.2
Space complexity	6.5.2
Space division multiplexing	3.4.1.3.1
Spanning subgraph	5.2.2.1.6
Spanning tree	5.2.2.4.4
Sparse matrix	5.1.4.1.5
Special functions of applied mathematics	7.1.2.5.1.4
Special purpose computer system	2.3
Special purpose controller	8.2.1.2.1
Special purpose simulation language	7.7.3.3.1.3
Specification of programs	4.1.3.2 4.1.3.7.1
Speech production	7.2.4.6 8.3.1.3.5
Speech recognition	1.4.1.7 7.2.4.1
	7.6.3.1.2 8.3.1.3.4
Speech response	7.2.4.6 1.4.1.7
Speed of a computer	1.1.1.1-5
Speed-up properties of algorithms	6.5.4.2
Spelling correction	8.4.3.4 8.4.3
Spline functions	5.1.6.1.5 8.2.2.4.2
Stability of numerical techniques	5.1.3.4 5.1.4.2.4
Stack	3.1.2.3.1 4.1.1.4.2 3.3.1.1
Stack automata	6.1.5.3
Standard deviation	8.2.2.4.5
Standards (u)	4.1.1.1.3 3.4.3.3
Star computer	9.3.3.3.3.3
Star-height-restricted language	6.2.2.3
State reduction	6.1.2.2
State trajectories	6.3.2.2.2

State transition	6.1.4.3
Statement (c)	**4.1.1.4** 4.1.1.4.1.1-2
	4.1.1.4.1.4 4.1.1.4.1.6
Statement-oriented simulation language	**7.7.2.3.2.1**
Static redundancy	**1.7.3.2.1**
Statistical data analysis	**8.2.2.3**
Steady-state behavior of a simulated system	**7.7.3.2.3**
Statistical feature extraction (IP)	**7.6.1.1.2.1**
Statistics re computer industry	**9.1.1**
Steepest descent	5.1.7.2.3 5.2.4.4.5
Stepwise refinement	**4.1.3.3.3**
	See also structured programming (u)
Stibitz relay computer	**9.3.2.2.2**
Stiff equations	**5.1.3.6**
Stirling number	**5.2.3.1.3**
Stochastic pattern recognition models	**7.6.1.1.2**
Stockholder records	**8.1.7.2**
Storage allocation	**4.2.1.3.2 7.8.5.3**
Storage location	4.1.1 See also cell (c)
Storage management	**4.1.1.4.2 7.4.5.2.2**
Storage media	**4.3.1.1.1**
Storage structure	**3.2**
Storage unit	1.2.1.2.2 1.2.2.1-4
Stored program computer	**9.3.2.3.6**
Store-and-forward message switching	**3.4.2.2**
Strassen's matrix multiplication algorithm	**6.6.1.4.1**
Stratified sampling	**8.2.2.3.3**
String	**3.1.1.5 4.1.1.5.1.4**
	7.8.1.2.1.1 6.1.3.3 6.3.1
String grammar	**7.3.1.4.2**
String homomorphism	6.2.3.4
Strip chart	**8.2.1.1.1**
Strips (AI)	**7.2.10.2**
Structure (software)	**3.1.2.7**
Structure based computer system	**2.1**
Structured programming (u)	
Stuck-at fault	**1.7.2.2.1.1**
Subroutine (c)	**4.1.1.4.1.5 4.1.2.1** 7.2.1.3
Subschema (DBMS)	**7.4.4.1.2**
Subscripted assignment	**6.3.3.1.1.2**
Substitution in algebraic expressions	**7.1.3.4.1**
Substitution cipher	**3.3.4.2**
Successive approximation	**5.1.5.1**
Summation method of image reconstruction	**7.5.5.1**
Summation of rational functions	**7.1.2.6**
Supercomputer	**1.1.1.5**
Superposition methods	**5.1.3.7.3**
Suppliers of computer products and services	**9.1.3**
Surface representation (CG)	**7.9.7.2**
Switching circuit	1.6.6 6.1
Switching theory	**6.1.1**
Symbol manipulation (u)	**7.1.3 8.2.4.2.3**
Symbolic algebra	See above
Symbolic name	4.1.1.3.2

Symmetric circuit	**1.5.1.1.2**
Synchronization statement	4.1.1.4.1.6
Synchronous time division multiplexing	3.4.1.3.3.1
Synonym dictionary (IR)	7.3.1.3.1
Syntactic representation	7.4.2.2
Syntax	4.1.1.1.2 4.1.1.3.5
	4.1.3.7.2 6.3.1
Syntax directed compiler	**4.1.1.2.2.1**
System development	**9.5.2 9.5.2.2**
System testing	**9.5.2.2.4**
Systems analysis (u)	**9.6.1.3**
Systems of equations	**5.1.1.6**
Systems programming	See applications programming (u)
Tablet input device	7.1.3.5.3
Table/questionnaire type languages	**4.1.1.3.6**
Tabular composition	8.4.2.2.2
Tagged data item	3.3.2.2.3
Tape cartridge	1.4.3.1.2
Tape cassette	1.4.3.1.1
Tape merge	7.8.1.2.2
Tape reel	1.4.3.1.3
Tape reversal complexity measure	6.5.3.1
Task	See job (c)
Taxation	**9.4.4**
Taylor series	**5.1.3.1.1** 7.1.1.2.5
Telephone network control	2.3.4.5
Teleprocessing systems	2.2.5
Television monitor	1.4.3.4.2
Teletype	1.4.2.1.1
Tensor	7.1.4.1.1
Term association (IR)	7.3.1.1.6
Term extraction (IR)	7.3.1.1.1
Term weighting (IR)	7.3.1.1.2
Terminal	1.4.2 9.1.3.1.3
	1.4.4.3 8.3.1.2 **9.4.8.3.2**
Terminal symbol	6.2.1.4
Termination logics	**6.3.4.2**
Tesselation automata	6.4.3
Test generation	1.7.4
Testing of software	4.1.3.5.1
Texas Instruments microprocessor	9.3.3.3.1.6
Text editing	**7.3.6.2.2 8.4.1**
	2.3.4.3 8.4.2.2.4 4.2.2.6
Textural properties of images	7.5.7.2 7.9.7.3
Theorem proving	7.2.2
Theory of computation	**6.**
Thermal printer	1.4.2.1.3
Thesaurus (IR)	7.3.1.1.4
Thickness of a graph	5.2.2.3.4
Third generation computers	9.3.3.3
Third normal form (DBMS)	7.4.1.2
Three dimensional graphics	7.9.7 .
Threshold circuit	**1.5.1.1.3**

Thresholding (IP)	7.5.6.1
Throughput (c)	2.4.1.1
Tightly-coupled computer system	2.1.2.2.1
Time advance during simulation	7.7.2.2.6 7.7.3.2.5
Time complexity	6.5.1
Time division multiplexing	3.4.1.3.3 3.4.2.1.2
Time sharing	2.2.3.1 8.4.1.2.3
	9.1.3.3.2 9.3.3.3.2.2
Time vs. space complexity analysis	6.5.2.4
Top-down design	4.1.3.3.3
	See also structured programming (u)
Topological program properties	6.3.5.1
Tournament sort	7.8.1.1.1.2
Tracking (IP)	7.5.6.5
Trade secrets	9.4.1.3
Training of personnel	8.2.3.3 9.1.3.5.2 9.5.2.3.2
Transaction-based system	2.3.4.6 8.1.1 2.2.3.2
Transaction-oriented simulation language	7.7.2.3.1.1
Transcendental functions	7.1.1.2.4
Transformational grammmar	7.3.1.4.4
Transient behavior of a simulated system	7.7.2.2.2 7.7.3.2.2
Transient fault	1.7.2.2.2
Transistorized computers, early	9.3.3.2.1
Transitive closure	5.2.2.2.6 6.2.4.2
Translation of programs	4.1.1.2
Transmission media	3.4.1.1
Transposition cipher	3.3.4.1
Trapezoidal rule	5.1.2.2.1 5.1.2.3
Traveling salesman problem	5.2.2.2.5 6.6.3.2.4
Tree	5.2.2.4 3.1.2.5 6.3.1
	7.8.2.2.3 7.2.3.1 7.4.2.1.1
Tree automata	6.1.4.4
Tree based comparative search	7.8.2.2
Tree insertion sort	7.8.1.1.2.4
Tree selection sort	7.8.1.1.1.2
Tridiagonal matrix	5.2.4.5.3
Trie	3.1.2.5.2.1 7.8.2.2.3
Trigonometric functions	5.1.6.1.6 7.1.2.5.1.2
Truncated index search	7.8.2.3.1
Truncation	5.2.1.1
Trunk line	3.4.2.1
Tuple (DBMS)	7.4.3.4
Turing machines	6.1.5.4 6.2.1.1 6.4.1
Turnaround time (c)	
Tutorial programs	8.3.3.3 7.2.9.2
TV display device	1.4.3.4.2
Twos complement number system	3.2.1.1.1.2
Two-way automata	6.1.4.1
Type 0 grammar	6.2.1.1
Type 1 grammar	6.2.1.2
Type 2 grammar	6.2.1.3
Type 3 grammar	6.2.1.4
Typesetting	8.4.2.2.4
Typography	8.4.2.1

Unbalanced tape merge	**7.8.1.2.2.2**
Unconditional transfer of control	**4.1.1.4.1.4** See also branch (c)
Undecidable propositions	6.3.6.2
Undergraduate education	**9.2.1.2**
Undirected graph	**5.2.2.1.2 3.1.2.6.2** 5.2.2.4.2
Uniform distribution	8.2.2.2
Union	6.2.2.2 6.2.3.4 5.2.2.1.3
Unipolar semiconductor element	1.6.1.2
Univac 80/90	9.3.3.1.2
Univac 1100 series	9.3.3.1.3 9.3.3.2.2 9.3.3.3.1.3
Univac 1	9.3.3.1.1 9.3.3.1.4.2
Universal instantiation (AI)	7.2.2.1
Universal logic circuit	**1.5.1.4.7**
Unordered file	**3.2.3.2.1**
Unrestricted grammars	**6.2.1.1**
Unscoped assignment	**6.3.3.1.1**
Unsharp masking (IP)	**7.5.3.4.2**
Unsolvability	**6.6.6**
US E users' group	9.6.2.2.3
User services	9.5.4
Utility programs	**4.2.2**
Vacuum tube	9.3.2.3.1
Valiant's parsing reduction algorithm	6.2.4.2
Variable (c)	4.1.1.3.3 4.2.2.1 6.1.1.1
Variable length record	3.2.2.2 3.3.2.2
Variable precision floating point number	7.1.1.4.5
Variational method	5.1.3.7.6
Vaucanson calculator	9.3.1.2.3
Vector	3.1.2.2
Vector display device	7.9.1.1.1.1
Vector processor	1.3.2.1.5 9.3.3.3.3.3
Vehicle routing	8.1.4.1
Verification of data	4.2.3.2.1
Verification of program correctness	4.1.3.7.2 5.3.2.4 6.3.4 7.1.4.3.2 7.2.1.2 4.1.3.3.5 7.4.1.5
Vertex of a graph	5.2.2
Vienna definition language (VDL)	4.1.1.1.2 6.3.2.1.2
Viewing operations (CG)	**7.9.3.2** 7.9.1.1.7.4
VI M users' group	9.6.2.2.4
Virtual processor	6.3.3.2.2
Virtual storage	**4.2.1.3.2.2 9.3.3.2.3** 7.8.1.1
Vision (AI)	**7.2.8**
Vocabulary display (IR)	7.3.3.4.1
Von Neumann, John	9.3.2.3.5
Voting	9.7.4.4.1
Wang's variant of a Turing machine	6.4.1.1.3
Waveform analysis	7.6.3.1
Weather prediction	8.2.3.2
Weight balanced tree	3.1.2.5.1.3
Whirlwind	9.3.2.3.6

Wiener filtering (IP)	**7.5.4.2**
Window (CG)	**7.9.3.2.1** 7.9.1.1.7.4
Word algebra	6.3.1.2
Word length (c)	1.1.1.1-5
Word processing	**8.4 2.3.4.3 9.5.3.3.2**
Word stems (IR)	**7.3.1.3.1** 7.3.1.1.3
Word truncation (IR)	**7.3.1.1.3**
World-view of a simulation language	**7.7.2.3.1**
Worst-case algorithm analysis	**7.1.2.7.1**
Xerographic printer	1.4.2.1.3
Xilog microprocessor	9.3.3.3.1.6
Younger's parsing algorithm	6.2.4.2
Zero of an equation	**5.1.1** 7.1.1.4.3
Zooming (CG)	**7.9.6.4.2**
Zuse Z3 computer	**9.3.2.2.3**